LEARNING BY DISCOVERY
A Lab Manual for Calculus

Anita E. Solow, Editor

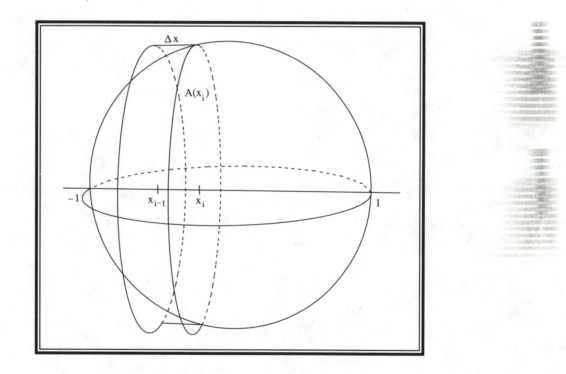

Volume 1

LEARNING BY DISCOVERY
A Lab Manual for Calculus

Anita E. Solow, Editor

A Project of
The Associated Colleges of the Midwest and
The Great Lakes Colleges Association

Writing Team

John B. Fink
Kalamazoo College

Bonnie Gold
Wabash College

Robert A. Messer
Albion College

Edward W. Packel
Lake Forest College

Supported by the National Science Foundation
A. Wayne Roberts, Project Director

MAA Notes Volume 27

Published and Distributed by
The Mathematical Association of America

MAA Notes and Reports Series

The MAA Notes and Reports Series, started in 1982, addresses a broad range of topics and themes of interest to all who are involved with undergraduate mathematics. The volumes in this series are readable, informative, and useful, and help the mathematical community keep up with developments of importance to mathematics.

MAA Notes

1. Problem Solving in the Mathematics Curriculum, *Committee on the Teaching of Undergraduate Mathematics,* a subcommittee of the Committee on the Undergraduate Program in Mathematics, *Alan H. Schoenfeld,* Editor

2. Recommendations on the Mathematical Preparation of Teachers, *Committee on the Undergraduate Program in Mathematics, Panel on Teacher Training.*

3. Undergraduate Mathematics Education in the People's Republic of China, *Lynn A. Steen,* Editor.

5. American Perspectives on the Fifth International Congress on Mathematical Education, *Warren Page,* Editor.

6. Toward a Lean and Lively Calculus, *Ronald G. Douglas,* Editor.

8. Calculus for a New Century, *Lynn A. Steen,* Editor.

9. Computers and Mathematics: The Use of Computers in Undergraduate Instruction, *Committee on Computers in Mathematics Education, D. A. Smith, G. J. Porter, L. C. Leinbach, and R. H. Wenger,* Editors.

10. Guidelines for the Continuing Mathematical Education of Teachers, *Committee on the Mathematical Education of Teachers.*

11. Keys to Improved Instruction by Teaching Assistants and Part-Time Instructors, *Committee on Teaching Assistants and Part-Time Instructors, Bettye Anne Case,* Editor.

13. Reshaping College Mathematics, *Committee on the Undergraduate Program in Mathematics, Lynn A. Steen,* Editor.

14. Mathematical Writing, by *Donald E. Knuth, Tracy Larrabee, and Paul M. Roberts.*

15. Discrete Mathematics in the First Two Years, *Anthony Ralston,* Editor.

16. Using Writing to Teach Mathematics, *Andrew Sterrett,* Editor.

17. Priming the Calculus Pump: Innovations and Resources, *Committee on Calculus Reform and the First Two Years,* a subcommittee of the Committee on the Undergraduate Program in Mathematics, *Thomas W. Tucker,* Editor.

18. Models for Undergraduate Research in Mathematics, *Lester Senechal,* Editor.

19. Visualization in Teaching and Learning Mathematics, *Committee on Computers in Mathematics Education, Steve Cunningham and Walter S. Zimmermann,* Editors.

20. The Laboratory Approach to Teaching Calculus, *L. Carl Leinbach et al.,* Editors.

21. Perspectives on Contemporary Statistics, *David C. Hoaglin and David S. Moore,* Editors.

22. Heeding the Call for Change: Suggestions for Curricular Action, *Lynn A. Steen,* Editor.

23. Statistical Abstract of Undergraduate Programs in the Mathematical Sciences and Computer Science in the United States: 1990–91 CBMS Survey, *Donald J. Albers, Don O. Loftsgaarden, Donald C. Rung, and Ann E. Watkins.*

24. Symbolic Computation in Undergraduate Mathematics Education, *Zaven A. Karian,* Editor.

25. The Concept of Function: Aspects of Epistemology and Pedagogy, *Guershon Harel and Ed Dubinsky,* Editors.

26. Statistics for the Twenty-First Century, *Florence and Sheldon Gordon*, Editors.

27. Resources for Calculus Collection, Volume 1: Learning by Discovery: A Lab Manual for Calculus, *Anita E. Solow*, Editor.

28. Resources for Calculus Collection, Volume 2: Calculus Problems for a New Century, *Robert Fraga*, Editor.

29. Resources for Calculus Collection, Volume 3: Applications of Calculus, *Philip Straffin*, Editor.

30. Resources for Calculus Collection, Volume 4: Problems for Student Investigation, *Michael B. Jackson and John R. Ramsay*, Editors.

31. Resources for Calculus Collection, Volume 5: Readings for Calculus, *Underwood Dudley*, Editor.

MAA Reports

1. A Curriculum in Flux: Mathematics at Two-Year Colleges, *Subcommittee on Mathematics Curriculum at Two-Year Colleges*, a joint committee of the MAA and the American Mathematical Association of Two-Year Colleges, *Ronald M. Davis*, Editor.

2. A Source Book for College Mathematics Teaching, *Committee on the Teaching of Undergraduate Mathematics*, *Alan H. Schoenfeld*, Editor.

3. A Call for Change: Recommendations for the Mathematical Preparation of Teachers of Mathematics, *Committee on the Mathematical Education of Teachers*, *James R. C. Leitzel*, Editor.

4. Library Recommendations for Undergraduate Mathematics, *CUPM ad hoc Subcommittee*, *Lynn A. Steen*, Editor.

5. Two-Year College Mathematics Library Recommendations, *CUPM ad hoc Subcommittee*, *Lynn A. Steen*, Editor.

These volumes may be ordered from the Mathematical Association of America, 1529 Eighteenth Street, NW, Washington, DC 20036.
202-387-5200   FAX 202-265-2384

Teachers may reproduce these modules for their students and they may be modified to suit particular classroom needs. However, the modules remain the property of the Mathematical Association of America and the collection may not be used for commercial gain.

First Printing

© 1993 by the Mathematical Association of America

ISBN 0-88385-083-4

Library of Congress Catalog Number 92-62279

Printed in the United States of America

Current Printing

10 9 8 7 6 5 4 3 2 1

INTRODUCTION
RESOURCES FOR CALCULUS COLLECTION

Beginning with a conference at Tulane University in January, 1986, there developed in the mathematics community a sense that calculus was not being taught in a way befitting a subject that was at once the culmination of the secondary mathematics curriculum and the gateway to collegiate science and mathematics. Far too many of the students who started the course were failing to complete it with a grade of C or better, and perhaps worse, an embarrassing number who did complete it professed either not to understand it or not to like it, or both. For most students it was not a satisfying culmination of their secondary preparation, and it was not a gateway to future work. It was an exit.

Much of the difficulty had to do with the delivery system: classes that were too large, senior faculty who had largely deserted the course, and teaching assistants whose time and interest were focused on their own graduate work. Other difficulties came from well intentioned efforts to pack into the course all the topics demanded by the increasing number of disciplines requiring calculus of their students. It was acknowledged, however, that if the course had indeed become a blur for students, it just might be because those choosing the topics to be presented and the methods for presenting them had not kept their goals in focus.

It was to these latter concerns that we responded in designing our project. We agreed that there ought to be an opportunity for students to discover instead of always being told. We agreed that the availability of calculators and computers not only called for exercises that would not be rendered trivial by such technology, but would in fact direct attention more to ideas than to techniques. It seemed to us that there should be explanations of applications of calculus that were self-contained, and both accessible and relevant to students. We were persuaded that calculus students should, like students in any other college course, have some assignments that called for library work, some pondering, some imagination, and above all, a clearly reasoned and written conclusion. Finally, we came to believe that there should be available to students some collateral readings that would set calculus in an intellectual context.

We reasoned that the achievement of these goals called for the availability of new materials, and that the uncertainty of just what might work, coupled with the number of people trying to address the difficulties, called for a large collection of materials from which individuals could select. Our goal was to develop such materials, and to encourage people to use them in any way they saw fit. In this spirit, and with the help of the Notes editor and committee of the Mathematical Association of America, we have produced five volumes of materials that are, with the exception of volume V where we do not hold original copyrights, meant to be in the public domain.

We expect that some of these materials may be copied directly and handed to an entire class, while others may be given to a single student or group of students. Some will provide a basis from which local adaptations can be developed. We will be pleased if authors ask for permission, which we expect to be generous in granting, to incorporate our materials into texts or laboratory manuals. We hope that in all of these ways, indeed in any way short of reproducing substantial segments to sell for profit, our material will be used to greatly expand ideas about how the calculus might be taught.

Though I as Project Director never entertained the idea that we could write a single text that would be acceptable to all 26 schools in the project, it was clear that some common notion of topics essential to any calculus course would be necessary to give us direction. The task of forging a common syllabus was managed by Andy Sterrett with a tact and efficiency that was instructive to us all, and the product of this work, an annotated core syllabus, appears as an appendix in Volume 1. Some of the other volumes refer to this syllabus to indicate where, in a course, certain materials might be used.

This project was situated in two consortia of liberal arts colleges, not because we intended to develop materials for this specific audience, but because our schools provide a large reservoir of classroom teachers who lavish on calculus the same attention a graduate faculty might give to its introductory analysis course. Our schools, in their totality, were equipped with most varieties of computer labs, and we included in our consortia many people who had become national leaders in the use of computer algebra systems.

We also felt that our campuses gave us the capability to test materials in the classroom. The size of our schools enables us to implement a new idea without cutting through the red tape of a larger institution, and we can just as quickly reverse ourselves when it is apparent that what we are doing is not working. We are practiced in going in both directions. Continual testing of the materials we were developing was seen as an integral part of our project, an activity that George Andrews, with the title of Project Evaluator, kept before us throughout the project.

The value of our contributions will now be judged by the larger mathematical community, but I was right in thinking that I could find in our consortia the great abundance of talent necessary for an undertaking of this magnitude. Anita Solow brought to the project a background of editorial work and quickly became not only one of the editors of our publications, but also a person to whom I turned for advice regarding the project as a whole. Phil Straffin, drawing on his association with UMAP, was an ideal person to edit a collection of applications, and was another person who brought editorial experience to our project. Woody Dudley came to the project as a writer well known for his witty and incisive commentary on mathematical literature, and was an ideal choice to assemble a collection of readings.

Our two editors least experienced in mathematical exposition, Bob Fraga and Mic Jackson, both justified the confidence we placed in them. They brought to the project an enthusiasm and freshness from which we all benefited, and they were able at all points in the project to draw upon an excellent corps of gifted and experienced writers. When, in the last months of the project, Mic Jackson took an overseas assignment on an Earlham program, it was possible to move John Ramsay into Mic's position precisely because of the excellent working relationship that had existed on these writing teams.

The entire team of five editors, project evaluator and syllabus coordinator worked together as a harmonious team over the five year duration of this project. Each member, in turn, developed a group of writers, readers, and classroom users as necessary to complete the task. I believe my chief contribution was to identify and bring these talented people together, and to see that they were supported both financially and by the human resources available in the schools that make up two remarkable consortia.

A. Wayne Roberts
Macalester College
1993

THE FIVE VOLUMES OF THE RESOURCES FOR CALCULUS COLLECTION

1. Learning by Discovery: A Lab Manual for Calculus
Anita E. Solow, editor

The availability of electronic aids for calculating makes it possible for students, led by good questions and suggested experiments, to discover for themselves numerous ideas once accessible only on the basis of theoretical considerations. This collection provides questions and suggestions on 26 different topics. Developed to be independent of any particular hardware or software, these materials can be the basis of formal computer labs or homework assignments. Although designed to be done with the help of a computer algebra system, most of the labs can be successfully done with a graphing calculator.

2. Calculus Problems for a New Century
Robert Fraga, editor

Students still need drill problems to help them master ideas and to give them a sense of progress in their studies. A calculator can be used in many cases, however, to render trivial a list of traditional exercises. This collection, organized by topics commonly grouped in sections of a traditional text, seeks to provide exercises that will accomplish the purposes mentioned above, even for the student making intelligent use of technology.

3. Applications of Calculus
Philip Straffin, editor

Everyone agrees that there should be available some self-contained examples of applications of the calculus that are tractable, relevant, and interesting to students. Here they are, 18 in number, in a form to be consulted by a teacher wanting to enrich a course, to be handed out to a class if it is deemed appropriate to take a day or two of class time for a good application, or to be handed to an individual student with interests not being covered in class.

4. Problems for Student Investigation
Michael B. Jackson and John R. Ramsay, editors

Calculus students should be expected to work on problems that require imagination, outside reading and consultation, cooperation, and coherent writing. They should work on open-ended problems that admit several different approaches and call upon students to defend both their methodology and their conclusion. Here is a source of 30 such projects.

5. Readings for Calculus
Underwood Dudley, editor

Faculty members in most disciplines provide students in beginning courses with some history of their subject, some sense not only of what was done by whom, but also of how the discipline has contributed to intellectual history. These essays, appropriate for duplicating and handing out as collateral reading aim to provide such background, and also to develop an understanding of how mathematicians view their discipline.

ACKNOWLEDGEMENTS

Besides serving as editors of the collections with which their names are associated, Underwood Dudley, Bob Fraga, Mic Jackson, John Ramsay, Anita Solow, and Phil Straffin joined George Andrews (Project Evaluator), Andy Sterrett (Syllabus Coordinator) and Wayne Roberts (Project Director) to form a steering committee. The activities of this group, together with the writers' groups assembled by the editors, were supported by two grants from the National Science Foundation.

The NSF grants also funded two conferences at Lake Forest College that were essential to getting wide participation in the consortia colleges, and enabled member colleges to integrate our materials into their courses.

The projects benefited greatly from the counsel of an Advisory Committee that consisted of Morton Brown, Creighton Buck, Jean Callaway, John Rigden, Truman Schwartz, George Sell, and Lynn Steen.

Macalester College served as the grant institution and fiscal agent for this project on behalf of the schools of the Associated Colleges of the Midwest (ACM) and Great Lakes Colleges Association (GLCA) listed below.

ACM	GLCA
Beloit College	Albion College
Carleton College	Antioch College
Coe College	Denison University
Colorado College	DePauw University
Cornell College	Earlham College
Grinnell College	Hope College
Knox College	Kalamazoo College
Lake Forest College	Kenyon College
Lawrence University	Oberlin College
Macalester College	Ohio Wesleyan University
Monmouth College	Wabash College
Ripon College	College of Wooster
St. Olaf College	
University of Chicago	

I would also like to thank Stan Wagon of Macalester College for providing the cover image for each volume in the collection.

Table of Contents

Preface

Welcome to *Learning by Discovery: A Lab Manual for Calculus*. These twenty-six labs cover topics from the entire spectrum of concepts usually contained in a calculus course including differential calculus, integral calculus, sequences and series, differential equations, and multivariable calculus.

Traditionally, the central ideas of calculus have been presented by teachers lecturing to their students. Although this is an efficient method for covering the syllabus, we feel that too much is lost. Students can learn a great deal by experimenting and discovering many of these important ideas themselves. It is the advent of technology that has made it possible for students, guided by leading questions, to discover for themselves things formerly accessible only by carefully orchestrated theoretical development. During the course of the labs, students are asked to create examples, to make conjectures, and sometimes even to prove results. In this way, students are actively involved in doing mathematics. For some, it is the first time they are excited by mathematics.

A word about the term *lab*. We are using lab, or laboratory module, to mean a set of exercises for the student. A lab could be done, for example, in a separate room full of computers, in a classroom with calculators, or in the student's own room using a computer or calculator. However it is done, technology is an integral part of these labs. It is used both to perform tedious calculations and to provide pictures to help students develop geometric insight. We feel this is a successful means of engaging students in a challenging learning experience. The authors have experience using these labs with students working with computers in a lab room. However, one reviewer reports having worked through most of this material using a graphics calculator.

We decided at the beginning of this project to write labs that were software-independent. This is different from most lab books and requires an explanation. The writing of this volume was supported by a grant from the National Science Foundation, and it was our intent to provide materials that could be useful to as wide a group of schools as possible and that would not be outdated by the rapid changes that are characteristic of software development. The twenty-six schools of the Associated Colleges of the Midwest and the Great Lakes Colleges Association gave us an ideal context in which to test ideas on a wide variety of machines.

You will find that these labs deal with the ideas of calculus, not with the details of technology. For example, there are no questions dealing with roundoff error. While we wanted to write materials that could be used with various software packages, we also wanted to make it easy for the instructor to use them. Therefore, at the end of each lab module we provide *Notes to Instructor* that discuss computing requirements and implementation of the labs in the three widely used computer algebra systems of Derive, Maple, and Mathematica. There are also some references for using graphing calculators.

Suggestions for Using these Materials

These labs can be used in a variety of ways. At many schools a computer lab session is held in place of a regularly scheduled class period. This can occur every week or on an irregular basis throughout the term. If you do not have the facilities for holding a lab session for your class, or if you do not wish to use class time, you can assign these labs as independent work outside of class.

Each laboratory module follows the same format. First the *Goals* of the lab are listed. Some labs have a section *Before the Lab* which contains non-computer exercises that give background for it. Students are to read through the instruction sheet and work out the material in this section prior to the lab period. The section *In the Lab* consists of problems, discussion, and hints for students to work on during the actual laboratory session. The labs end with some problems for *Further Exploration*. They are designed to get the students to think about the ideas and to try to apply them to slightly harder questions, usually without using the computer. The *Notes to Instructor* follow the lab instruction sheets. They contain the name of the principal author, scheduling suggestions, and computer or calculator requirements. We also mention ideas we had in mind when we wrote the labs, comments on particular questions in the labs, and comments on implementing the labs, including necessary computer programs.

Because these labs are software-independent, those of you using computers may need to supplement them with local computer instructions. These may take the form of a separate sheet of instructions (perhaps just relevant computer commands with brief comments as to what they do) to go along with each lab, or you may want to provide a reference sheet of commands for the entire semester. Sample computer programs for Derive, Maple, and Mathematica are given in the *Notes to Instructor* for several labs. We strongly suggest that you enter these programs into the software package before the lab and make them immediately available for the students to use.

The lab modules were written to allow for a great variety of styles of use. Below are some recommendations for their use. The recommendations are based on the experience of many instructors who use computers in teaching calculus.

- Before assigning a lab to your class, try it yourself using the same hardware and software that your students will be using. This experience will better inform you about the subtleties of the lab and give you some indication of the kinds of problems and learning experiences you can expect your students to have. A trial run will also give you a sense of how long the lab will take. If it is too long, you should decide what changes, hints, or omissions you will need to make for your class.

- A knowledgeable person must be available whenever students are working on a lab. The instructor would be best, at least for the first few labs. The students should be concentrating on the mathematical concepts of the labs and not on the syntax and peculiarities of a particular system. They therefore need quick assistance when something goes wrong.

- We recommend that students work in pairs. Working in pairs forces students to talk mathematics to each other and often enables them to solve problems that neither could do alone. Also, if both members of a pair are stymied, they are more likely to come for help promptly than if they were working alone. Three students working together is not recommended, however. Too often two sit and chat while the third does the work.

- Students should record their data and observations on clean sheets of paper or in a notebook. Scribbling numbers in the margins of the lab sheet inevitably leads to confusion. Discourage your students from mindlessly printing all graphs and data that the computer has generated. This practice wastes time, since the printers are much slower than the computers, and discourages thinking. Students should be encouraged to consider the essential aspects of the graph or data and to record only them in their notebooks. Often a graph sketched by hand will capture all that the student needs for the lab.

- Give your students copies of the lab instruction sheet well in advance of the lab. Instruct them to read it thoroughly and to complete any *Before the Lab* problems. Because time is often limited, students must be ready to do the lab when they enter the lab room.

- The lab experiences are most successful when they are integrated into the regular classroom work. You should discuss any necessary ideas when you hand out the instruction sheets in a class period before the lab, discuss the lab in subsequent classes, refer to the ideas of the lab in class as often as is practical, and include questions on the examinations that refer to the lab.

- Many of the lab problems ask the students to derive conclusions based on data they have produced in the lab. For this to succeed, the environment in the lab and classroom should allow the students to feel comfortable to make mistakes. Otherwise, they will not feel free enough to make the requested conjectures.

Student lab reports can be handled in many ways. One common approach is to have the students hand in the lab report one week after the lab session. In this case, the reports should be fairly substantial and well written. Some of us encourage the lab partners to hand in one report; others insist on separate ones. Another idea is to have the lab report, except for the answers to the *Further Exploration* questions, due at the next class, where the main ideas of the lab will be used. This may be particularly useful for those labs that are intended to have students discover some of the major results of calculus. Some or all of the *Further Exploration* problems can then be assigned for the following week. Some instructors have their students write the lab reports electronically at the same time they are taking the lab. Others have only a few lab reports due during the semester, insisting that each be a major project. In this model it is less appropriate for students to answer numbered questions than for them to write papers that discuss the findings of the lab.

Calculus laboratories are new for most of us, and using this approach has a great impact on the time that we spend teaching. We have provided these labs so you do not need to write them yourself. Unfortunately, we cannot remove the burden of grading lab reports. Yet, if you expect your students to take the labs seriously and to view them as an important part of the course, then the reports need to be read and graded, and the grades must be a significant part of the course grade.

One practical method of reducing the time spent grading is to have each pair of students hand in a single report. Another is to collect the data sheets at the end of the lab period. These can be quickly checked and returned to the students the next day. Students can then write a short report summarizing the ideas of the lab. Some instructors find that if they fix the format of the lab report so they know where to look for each answer, grading becomes much quicker and easier. After the reports are graded, you may want to choose the best report as an answer key for the class. This provides an incentive for students to do well and saves the time it would take for you to write up a sample lab report.

Following this preface, you will find *A Note to Students*. We suggest that you distribute a copy of this document, or your own version, to all students. The demands of calculus labs are often quite different from the demands that have traditionally been placed on students in mathematics courses. Therefore, we have found that it is helpful to spell out our expectations for students doing labs. We need to tell them what they should do before the lab in order to be prepared, what they should do in the lab, and how to report their findings.

The lab modules in this volume have been classroom tested during the past three years, and they have been revised in response to suggestions from the instructors and students who used them. Several reviewers also offered helpful comments. In particular, I wish to thank Robert Eslinger (Hendrix College) and Creighton Buck (University of Wisconsin) for their detailed and careful readings of the manuscript. Robert Messer, in addition to his role as a writer, designed the format for this volume. Although each lab lists a principal author, the writing and editing was a collaborative effort by the entire writing team. I found them to be a multi-talented group who were a delight to work with, and I thank them all for their devotion to the project.

> Anita E. Solow
> Grinnell College
> 1992

A Note to Students

For many of you a mathematics lab, especially one that introduces new ideas and encourages "learning by discovery," will be a new experience. So that you can take full advantage of this activity, we offer the following suggestions.

The Laboratory Session

- Read over the laboratory instructions carefully before the sessions starts.

- If work is called for *Before the Lab*, make sure to do it with understanding. A successful lab experience will depend upon this work.

- Think as you work through the steps of the lab. Always ask yourself and your partner if the output makes sense.

- Discriminate between important and incidental computer output. Record the important output, along with your thoughts and observations, on a data sheet or in a notebook. The data sheet will be needed when you write your lab report and may sometimes be collected by your instructor.

The Lab Report

The lab report should be a thoughtful, well-written, and neatly organized document that summarizes both your experience in the lab and what you learned as a result of that experience. Your report should contain the following four parts.

1. *Heading.* At the top list the title of the lab, your name, and the name of the other student who worked with you on the lab.

2. *Data.* Summarize the data you collect in a succinct, easy-to-grasp form, such as a table or a picture with labels. Keep in mind that your instructor is interested in your answers, thoughts, and analysis, rather than in output from the computer.

3. *Calculations.* If the lab instruction sheet asks for calculations or graphs based on your data, present them succinctly. Explain briefly how you did your calculations, perhaps by writing out one typical calculation in full.

4. *Conclusions.* Write your conclusions in a paragraph or two. They should be inferences you draw from your data and calculations. Here is your opportunity to show that you understood the purpose of the lab, saw patterns in the data, and gained significant insights. Be as sweeping in your conclusions as you dare, but back them up by explicit references to your data and calculations.

Lab 1: Graphing Functions

Goals

- To become familiar with graphing functions on the computer.

- To explore the behavior of families of functions.

In the Lab

1. In this problem we will explore the effect that changing the coefficient of x has on the graph of the parabola $y = 3x^2 + bx - 2$.

   a. First consider positive values of b. Use the computer to graph $y = 3x^2 + x - 2$, $y = 3x^2 + 2x - 2$, and $y = 3x^2 + 5x - 2$ on the same set of axes. How does changing the value of b affect the vertex of $y = 3x^2 + bx - 2$? Feel free to try additional positive values of b to convince yourself that your answer is correct.

   b. Now graph $y = 3x^2 + bx - 2$ for $b = -1$, -2, and -4. When the coefficient b is negative, how does changing its value affect the parabola?

   c. Summarize the effect that the value of b has on the position of the vertex of the parabola. Make sure that your answer is valid for all values of b, positive, negative or zero.

   d. Notice that all the graphs have the same y-intercept. Give a simple explanation for this phenomenon.

2. a. Graph $y = \sin x$ for $-2\pi \le x \le 2\pi$. Graph the following four functions on the same domain.

   $$p(x) = \sin(2x) \qquad q(x) = 2\sin x \qquad r(x) = \sin x + 2 \qquad s(x) = \sin(x + 2)$$

   On your data sheet, draw a rough sketch of each of these functions, noting the period, amplitude, intercepts, and other relevant details.

   b. In each case explain the effect the number 2 had in modifying the graph of $y = \sin x$.

   c. Let $g(x) = x^3 - x$. Compare the graphs of $y = g(ax)$ and $y = ag(x)$ with the graph of $y = g(x)$ for $a = .5$, 3, and 4.

d. Let f be any function and let a be any positive constant. Make a general statement about the relation of the graphs of $y = f(ax)$ and $y = af(x)$ to the graph of $y = f(x)$. If you are not sure from your work above, feel free to experiment with other functions such as $f(x) = |x|$ or $f(x) = \sec x$.

e. Again, using $g(x) = x^3 - x$, compare the graphs of $y = g(x + a)$ and $y = g(x) + a$ with the graph of $y = g(x)$ for $a = .5$, 3, and -3.

f. Let f be any function and a be any constant. How are the graphs of $y = f(x + a)$ and $y = f(x) + a$ related to the graph of $y = f(x)$?

3. Let $f(x) = \dfrac{2x}{3x + 1}$. We will study the effect of the absolute value function on the graph of this function.

a. Graph the function f and discuss the main features of the graph, such as intercepts, and horizontal and vertical asymptotes.

b. Graph $y = |f(x)|$. How does the absolute value function affect the graph of f?

c. Graph $y = f(|x|)$ and discuss the relation of this graph to the graph of f.

d. In Figure 1, the graph of $y = h(x)$ is given. On this figure sketch the graph of $y = |h(x)|$. In another color, sketch the graph of $y = h(|x|)$. Give reasons for your sketches.

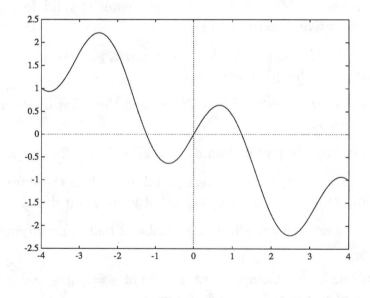

Figure 1

4. Consider the quadratic polynomial $kx^2 + (k+1)x - (k+2)$, where k is a constant.

    a. Graph the parabolas $y = x^2 + 2x - 3$ (where $k = 1$) and $y = 2x^2 + 3x - 4$ (where $k = 2$). Also graph the parabolas for $k = 3$, 4, and a few larger values. Compare the vertices and general shapes of these parabolas. What is happening as k increases?

    b. Notice that all of these parabolas cross the x-axis at two points. Let s_k be the larger of the two solutions of $kx^2 + (k+1)x - (k+2) = 0$. By looking at the graphs, approximate s_1, s_2, s_3, and s_4.

    c. Guess the behavior of s_k as k gets very large. Hint: You can always test your guess by graphing the polynomial for larger values of k.

Further Exploration

5. Complete the square in the formula for the parabola in Problem 1 to give an algebraic explanation of your observations.

6. Use the quadratic formula to find the roots of the polynomial in Problem 4. Use the formula for s_k to calculate $\lim_{k \to \infty} s_k$.

Notes to Instructor   Lab 1: Graphing Functions

Principal author: Bonnie Gold

Scheduling: This lab should be scheduled at the beginning of the first semester of calculus. Exploration problem 6 refers to limits.

Computer or calculator requirements: Ability to graph several functions on the same set of axes. Unless your computer system uses color or some other means of distinguishing multiple graphs on the same set of axes, students should not try to compare more than two or three functions on the same graph. Encourage them to sketch the results in their notebooks.

Comments: You will need to provide your students with an introduction to the calculator or computer system that they will be using. If possible, try to demonstrate its use in class before the first scheduled lab.

This lab has virtually no new mathematics in it. The goal is for students to become familiar with the computer and software that they will be using in subsequent labs. Although it covers only precalculus topics, students will need the graphs to answer most of the questions.

Problem 4 is due to Michael Henle (Oberlin College).

Lab 2: Introduction to Limits of Functions

Goals

- To develop an intuitive understanding of the nature of limits.

- To lay the foundation for the frequent use of limits in calculus.

- To experience the power and peril of investigating limits by successively closer evaluation.

In the Lab

In this lab we shall study the behavior of a function f near a specified point. While this is sometimes a straightforward process, it can also be quite subtle; in many instances in calculus the process for finding a limit must be applied carefully. By gaining an intuitive feel for the notion of limits, you will be laying a solid foundation for success in calculus.

1. Consider the function f defined by $f(x) = \dfrac{x^4 - 1}{x - 1}$.

   a. By successive evaluation of f at $x = 1.8$, 1.9, 1.99, 1.999, and 1.9999, what do you think happens to the values of f as x increases towards 2?

   b. Do a similar experiment on f for values of x slightly greater than 2. Again, comment on your results.

   As a shorthand, and anticipating a forthcoming definition, we shall describe what you found in parts a and b by writing $\lim\limits_{x \to 2} f(x) = 15$, or more specifically,

   $$\lim_{x \to 2} \frac{x^4 - 1}{x - 1} = 15.$$

   c. In this particular case you could have "cheated" by immediately evaluating f at 2. Get a computer plot of the function between 1.8 and 2.2 to illustrate what happens in this straightforward situation.

2. Use the same function f as above, but this time consider what happens as x approaches 1.

   a. Study this situation experimentally as you did in parts a and b of Problem 1. To gain some additional feel and respect for the situation, compute the numerator and denominator of f separately for several x values before dividing. What are your conclusions and, in particular, what is $\lim_{x \to 1} f(x)$?

   b. What happens when you try to "cheat" as was done in part c of Problem 1? There are situations in which direct evaluation at the specified point is possible and actually gives the limit. These give rise to a concept called *continuity*. There are many important situations in calculus when this technique will not work, however.

3. By computer or calculator experimentation, try to determine the values of the following limits. Use either graphing or function evaluation at nearby points.

   a. $\lim_{x \to 0} \dfrac{\sin(10x)}{x}$

   b. $\lim_{x \to 1} g(x)$ where
   $$g(x) = \begin{cases} \dfrac{x^4 - 1}{x - 1}, & \text{for } x < 1, \\ 17, & \text{for } x = 1, \\ 14 - 10/x, & \text{for } x > 1 \end{cases}$$

   Important Comment and Hint: As far as the existence and value of the limit are concerned, the value of $g(1)$ has no relevance.

   c. $\lim_{x \to 0} (1 + x)^{1/x}$

   Record your best guess; the limit is a famous mathematical constant.

   d. $\lim_{t \to 1} \dfrac{t^n - 1}{t - 1}$ for a general positive integer n

   Hint: Recall Problem 2, try other values of n, and generalize.

4. It is important to be aware that limits can sometimes fail to exist. Investigate the following limits and explain why you think each does not exist. You may find it helpful to use the computer to evaluate the functions at different values of x and to plot graphs of the functions.

   a. $\lim_{x \to 2} \dfrac{x}{x - 2}$

   b. $\lim_{x \to 0} \dfrac{\sin(10x)}{x^2}$

c. $\displaystyle\lim_{x \to 0} \sin(1/x)$

d. $\displaystyle\lim_{x \to 0} |x|/x$

e. $\displaystyle\lim_{x \to 4} f(x)$, where

$$f(x) = \begin{cases} (x+2)^3, & \text{for } x < 4, \\ e^x, & \text{for } x > 4 \end{cases}$$

Note: For parts *d* and *e* think about the idea of a "one-sided limit" and store your thoughts for future reference. You should also convince yourself that computer evaluation is not really needed to make wise conclusions in these particular situations.

Further Exploration

5. Find $\displaystyle\lim_{x \to \infty} f(x)$, if it exists, for $f(x) = \dfrac{3x^4 - x^2 + 10}{2x^4 + 5/x}$, $f(x) = \dfrac{\sin x}{1 + x^2}$, and $f(x) = \dfrac{4 - 3/x}{\sin x}$. Hints: The limit exists for two of these functions. Answer this question by using computer evaluation, common sense, and perhaps some algebra.

6. Try to determine the limit $\displaystyle\lim_{h \to 0} \dfrac{f(x+h) - f(x)}{h}$ when $f(x) = \ln x$. Hints: After specifying the quotient for the given f, treat x as a constant and give x a specific value of your own choosing before investigating the limit as h goes to 0. Then repeat for several other x values, make a table of results, and try to see the pattern. What function of x emerges as you compute this limit for values of x? Limits of this particular *difference quotient* are very important in calculus.

Notes to Instructor   Lab 2: Introduction to Limits of Functions

Principal author: Ed Packel

Scheduling: It is suggested that this lab be done just before the idea of limits of functions is formally introduced in class. The lab does use exponential and logarithmic functions, but knowledge of specific properties of these functions is not needed.

Computer or calculator requirements: Graph plotting and function evaluation.

Comments: Because of roundoff error, some computer algebra systems may give confusing results when ratios of small quantities are involved. Instructors may choose to identify such problems and give students advance warning or to let them occur as part of the laboratory learning experience.

While early parts of the lab call for extensive evaluation of functions near the point at which the limit is to be taken, students should be urged to minimize this approach in the latter parts of the lab, employing it only as a last resort. Thus, repeated evaluation may be necessary in Problems 2a and 2b, but should not be needed in 3b, for instance, to study $14 - 10/x$ for values of x just greater than 1.

If the computer software being used has a `limit` command, the instructor may want to tell students how to use it at some point near the end of the lab. This would be especially timely if the limit command will be used in later labs or exercises.

In Problem 3b it is not expected that the student should define the function g for the computer. A suggestion may be in order about how g should be evaluated by the student and by the computer for points near 1.

Do not let students get too bogged down on Problem 6. It is a some-what higher level "discovery" question and should be received in that spirit. To make the problem easier to tackle, it may be replaced by a question involving $\lim_{h\to 0} \frac{\ln(2+h) - \ln(2)}{h}$.

Comments on implementing the lab: To graph a piecewise defined function, see the instructions provided in *Notes to Instructor* following Lab 5.

Lab 3: Zooming In

Goals

- To define the slope of a function at a point by zooming in on that point.

- To see examples where the slope, and hence the derivative, is not defined.

In the Lab

You learned in analytic geometry that the slope of a non-vertical straight line is $\dfrac{\Delta y}{\Delta x}$ or $\dfrac{y_2 - y_1}{x_2 - x_1}$, where (x_1, y_1) and (x_2, y_2) are any two points on the line. Most functions we see in calculus have the property that if we pick a point on the graph of the function and zoom in, we will see a straight line.

1. Graph the function $f(x) = x^4 - 10x^2 + 3x$ for $-4 \le x \le 4$. Zoom in on the point $(3, 0)$ until the graph looks like a straight line. Pick a point on the curve other than the point $(3, 0)$ and estimate the coordinates of this point. Calculate the slope of the line through these two points.

The number computed above is an approximation to the *slope of the function* $f(x) = x^4 - 10x^2 + 3x$ *at the point* $(3, 0)$. This slope is also called the *derivative of* f *at* $x = 3$, and is denoted $f'(3)$.

2. Use zooming to estimate the slope of the following functions at the specified points.

   a. $f(x) = x^4 - 6x^2$ at $(1, -5)$

   b. $f(x) = \cos x$ at $(0, 1)$

   c. $f(x) = \cos x$ at $\left(\frac{\pi}{2}, 0\right)$

   d. $f(x) = (x - 1)^{1/3}$ at $(2, 1)$

So far in this lab you have used the graph of a function to estimate the value of its derivative at a specified point. Sometimes, however, a function does not have a slope at a point and therefore has no derivative there.

3. Graph $f(x) = (x - 1)^{1/3}$ again. This time, zoom in on $(1, 0)$. Describe what you see. By examining your graphs, explain why the slope is not a finite real number at $x = 1$. For this function, conclude that $f'(1)$ does not exist.

4. Graph the function $f(x) = |x^4 - 6x^2|$. By looking at the graph and zooming in on points you select, decide at which points the function f has a derivative and at which points it does not. Support your answers with appropriate sketches.

Further Exploration

5. The derivative of f at a point a is defined analytically by the formula

$$f'(a) = \lim_{h \to 0} \frac{f(a + h) - f(a)}{h}.$$

Explain in your own words how calculating the slope of a function at the point $(a, f(a))$ by repeated zooming is related to the computation of the derivative $f'(a)$ by this definition.

6. Use zooming to investigate $f'(0)$, the slope of the curve $y = f(x)$ at $x = 0$, for the following two functions.

    a.
$$f(x) = \begin{cases} x \sin(\frac{1}{x}), & \text{if } x \neq 0, \\ 0, & \text{if } x = 0 \end{cases}$$

    b.
$$f(x) = \begin{cases} x^2 \sin(\frac{1}{x}), & \text{if } x \neq 0, \\ 0, & \text{if } x = 0 \end{cases}$$

Notes to Instructor   Lab 3: Zooming In

Principal author: Anita Solow

Computer or calculator requirements: Ability to graph functions. It is helpful to be able to find the coordinates of points on a graph.

Scheduling: This lab is an introduction to the idea of the derivative and should be scheduled before derivatives are defined in class. It is an alternative to the more ambitious lab "Discovering the Derivative."

Comments: This lab introduces the derivative geometrically by zooming in on a smooth function at a specified point until the curve appears straight. The slope of this line is then defined to be an approximation to the slope of the curve at that point. The strength of this approach is that the idea of a derivative is based upon a concept with which students are familiar. This is in contrast with the standard definition which relies on limits. A more thorough discussion of this approach can be found in David Tall's article, "Intuition and Rigour: The Role of Visualization in the Calculus," in the MAA Notes Number 19, *Visualization in Teaching and Learning Mathematics,* edited by Walter Zimmermann and Steve Cunningham (1991). Another benefit of this approach is that it supports the idea of a linear approximation to a curve when it is introduced later in the course.

Comments on implementing the lab: Zooming is either accomplished by redefining the domain or by changing the scale. Your software may have a built-in zoom command for this purpose. In any case, keep the specified point in the center while you zoom. For example, in Problem 1, if you need to redefine the domain, try $[2, 4]$, $[2.5, 3.5]$, $[2.8, 3.2]$, $[2.9, 3.1]$, $[2.99, 3.01]$, etc.

In Derive, issue Manage-Branch-Real before the lab so that real values of $(x - 1)^{1/3}$ will be computed.

Lab 4: Discovering the Derivative

Goals

- To define the slope of a function at a point by zooming in on that point.

- To develop the definition of the derivative of a function at a point by examining slopes of secant lines.

- To understand situations in which the derivative will fail to exist.

In the Lab

You learned in analytic geometry that the slope of a non-vertical straight line is $\dfrac{\Delta y}{\Delta x}$ or $\dfrac{y_2 - y_1}{x_2 - x_1}$, where (x_1, y_1) and (x_2, y_2) are any two points on the line. Most functions we see in calculus have the property that if we pick a point on the graph of the function and zoom in, we will see a straight line.

1. Graph the function $f(x) = x^3 - 6x + 3$ for $-3 \leq x \leq 4$. Zoom in on the point $(1, -2)$ until the graph looks like a straight line. Pick another point on the curve other than $(1, -2)$ and estimate the coordinates of this point. Calculate the slope of the straight line through these two points.

   The number computed above is an approximation to the *slope of the function* $f(x) = x^3 - 6x + 3$ *at the point* $(1, -2)$. This slope is also called the *derivative of* f *at* $x = 1$, and is denoted by $f'(1)$.

   In Problem 1 you learned how to use the graph of a function to estimate the value of its derivative at a specified point. There is an analytic definition of the derivative that you will now develop using a primarily geometric approach.

2. In Figure 1, the straight line intersects the graph of the function f at two points with x-coordinates a and $a + h$. Write expressions for the coordinates of these two points and then write a formula for the slope of the straight line.

3. Again, consider the function f defined by $f(x) = x^3 - 6x + 3$, and let $a = 1$. Its graph over the domain $[0, 2]$ is given in Figure 2.

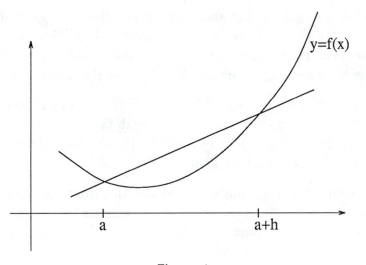

Figure 1

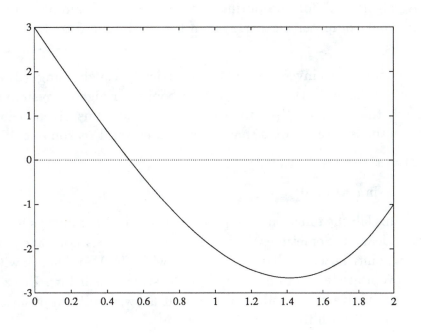

Figure 2

a. For each value of h, sketch by hand, on Figure 2, the straight line through $\left(1, f(1)\right)$ and $\left(1+h, f(1+h)\right)$. Make a table containing the values of h and the slopes of the straight lines. Use $h = 0.6, \quad 0.4, \quad 0.2, \quad -0.6, \quad -0.4$, and -0.2.

b. As the h values get smaller and smaller (positive or negative), what geometrical property do the lines we have drawn and the line to which they tend seem to have? By inspecting your graph and the slopes you have computed, try to guess a limiting value for the slopes of the straight lines.

c. Based on your answer to Problem 2 and your thoughts on parts a and b, write an expression (in terms of a general f) for the slope of the limiting (that's a hint) straight line referred to in b. We call this expression the *derivative of f at a*, and we denote it by $f'(a)$.

d. Use the computer to estimate or evaluate this limit for the specific function $f(x) = x^3 - 6x + 3$ and $a = 1$. Is the answer consistent with the slope estimate you made in part b?

We now have an analytic definition corresponding to the geometric picture of the derivative of a function at a point. It is an important aspect of calculus that there is an "algorithm" for computing the derivatives of familiar functions. Not surprisingly, your computer algebra system has a command to give values for the derivative $f'(a)$.

4. Use the computer to investigate (1) geometrically via zooming, (2) analytically via use of limits, and (3) directly via your computer algebra system's derivative command, the value of $f'(a)$ for each of the situations given below. Do the various methods give answers that are consistent and reasonable?

a. $f(x) = x^4 - x^2 + 3$ at $(1, 3)$

b. $f(x) = \sin x$ at $(\pi, 0)$

5. So far in this lab the functions and points considered have always had derivatives that were defined. Sometimes, however, a function does not have a well-defined slope at a point. Discuss what is happening with $f'(a)$, and hence with the slope of the appropriate tangent line, in each of the situations below. Use zooming and then the derivative command. Support your answers with appropriate sketches of the graphs of both functions.

a. $f(x) = (x - 1)^{1/3}$ at $a = 1$

b. $f(x) = |x^2 - 4|$ at a general a

    For what values of a does $f'(a)$ exist and for what values does it fail to exist?

Further Exploration

6. Explain in your own words how calculating the slope of a function at the point $(a, f(a))$ by repeated zooming is related to the definition of the derivative $f'(a)$ that you discovered in Problem 3c.

Notes to Instructor   Lab 4: Discovering the Derivative

Principal author: Ed Packel

Scheduling: This lab is an introduction to the idea of the derivative and should be scheduled before derivatives are defined in class. A shorter version concentrating on the idea of zooming is found in the alternative lab "Zooming In."

Computer or calculator requirements: Graph plotting, computing or estimating limits, symbolic differentiation.

Comments: This lab encourages the student to discover the limit definition of the derivative by looking at graphs of functions and secant lines that approach the tangent line to the graph at a point. For comments about the idea behind zooming, see the *Notes to Instructor* following Lab 3.

Comments on implementing the lab: See the *Notes to Instructor* at the end of "Zooming In."

Lab 5: Investigating the Intermediate Value Theorem

Goals

- To discover and acquire a feel for one of the major theorems in calculus.

- To apply the theorem in both practical and theoretical ways.

- To understand why continuity is required for the theorem.

In the Lab

This lab will motivate you to discover an important general theorem of calculus. The theorem, called the Intermediate Value Theorem (IVT for short—its name is a clue to what it says), has a clear geometric interpretation but requires a careful mathematical formulation. This lab will develop your intuition for both aspects of the theorem.

1. Consider the function f defined on the closed interval $[1, 2]$ by $f(x) = x^5 - 4x^2 + 1$.

   a. Compute the numerical values of $f(1)$ and $f(2)$. Explain why you think the graph of f must cross the x-axis somewhere between $x = 1$ and $x = 2$. That is, why must there be a number c between 1 and 2 such that $f(c) = 0$?

   b. Plot the graph of the function f on $[1, 2]$ to support what you said in part a above. Also, estimate any zeros of f (points c at which $f(c) = 0$) on $[1, 2]$.

2. Consider the function g defined on the closed interval $[-3, 2]$ by

   $$g(x) = \begin{cases} x^2 + 1, & \text{for } -3 \le x \le 0, \\ 1 - x^2 - x^4, & \text{for } 0 < x \le 2. \end{cases}$$

   a. Compute the numerical values of $g(-3)$ and $g(2)$. Explain why you think the graph of g must cross the x-axis somewhere between $x = -3$ and $x = 2$.

   b. Plot the graph of g on $[-3, 2]$ and estimate any zeros of g.

   c. Now replace g by the slightly altered function h defined on $[-3, 2]$ by

   $$h(x) = \begin{cases} x^2 + 1, & \text{for } -3 \le x \le 0, \\ -1 - x^2 - x^4, & \text{for } 0 < x \le 2. \end{cases}$$

Once again evaluate the function at its endpoints and think about whether
it will take on the value 0 somewhere in the interval. Also plot the graph,
and try to explain in general terms why you came to your conclusion.

3. We are now ready to formulate a statement of the Intermediate Value Theorem.
    Based upon the observations above, fill in the blanks to complete the following.

    Given a _____ function f defined on the closed interval
    $[a, b]$ for which 0 is between _____ and _____, there
    exists a point c between _____ and _____ such that
    _____ .

    Important Note: The above statement of the IVT is an example of what is called
    an "existence theorem." It says that a certain point exists, but does not give a
    rule or algorithm for how to find it. (Lovely algorithms for approximating such
    a point to any desired degree of accuracy do exist.) In the applications of the
    theorem, the key fact is that such a point exists, not its specific value.

Some Typical Applications

4. *Friendly warm-up.* Use the IVT to prove that the function defined by $f(x) = \sin x^2 - \cos x$ has a zero (that is, a point c at which $f(c) = 0$) on the interval
    $[0, 1]$. Use graphing and/or root finding methods to estimate a zero on the
    interval.

5. Use the IVT to prove that the function defined by $f(x) = x^3 - e^x + e^{-x}$ has
    at least five zeros on the real line. Hint: On $[-5, 5]$ find six points at which f
    has alternate signs. Then apply the IVT on five appropriately chosen intervals.
    Note that a single application of IVT on $[-5, 5]$ guarantees only one zero, but
    it does not rule out the possibility of more than one zero. Use graphing and/or
    root finding methods to estimate the five zeros of f.

6. If your oven is at 250° and you turn it off, is there ever an instant when the
    oven temperature is 170°? Explain your answer and its relation to the IVT.

7. If you remove marbles from a bag one at a time, must there always come a
    time when the bag contains exactly half the number of marbles it began with?
    Again, explain your answer and its relation to the IVT.

8. In your notebook sketch the graph of a continuous function g over the interval $[0, 1]$. Now draw the graph of another continuous function h over the same interval with the property that $h(0) < g(0)$ and $h(1) > g(1)$.

   a. Must these two graphs cross? Express this behavior in terms of a condition involving the functions g and h and a point c in the interval $(0, 1)$.

   b. Give a proof that what you observed in part a must always be true for any two continuous functions g and h on $[0, 1]$ with the property that $g(0) > h(0)$ and $h(1) > g(1)$. Hint: Apply the IVT thoughtfully to the function $f = g - h$.

   c. One plate has been in the freezer for a while, the other is in a warm oven. The locations of the two plates are then switched. Will there be a moment when the plates are at the same temperature? How does your answer relate to the ideas developed in parts a and b above?

   d. Show that on any circle (such as a great circle on the surface of the earth) there is always a pair of points at the opposite ends of a diameter that are at the same temperature.

9. *Fixed point theorem.* A *fixed point* of a function f is a point c in the domain of f for which $f(c) = c$.

   a. In your notebook, sketch the graph of a continuous function f defined on an interval $[a, b]$ and whose values also lie in the interval $[a, b]$. You may choose the endpoints a and b of the interval arbitrarily, but be sure that the image of the function is contained within its domain. Locate a point c on your graph that is a fixed point of f.

   b. Try to draw the graph of a continuous function f defined on an interval $[a, b]$ with values in the interval $[a, b]$ that has no fixed points. What is getting in your way? Prove that any continuous function f defined on $[a, b]$ with values in $[a, b]$ must have a fixed point c in the interval $[a, b]$. Hint: Use your pictures for guidance and draw in the graph of the line $y = x$. Ask your instructor for further hints if you are still stuck.

Further Exploration

10. *A more general IVT.* There is nothing special, as far as the IVT is concerned, about a function having the value 0.

   a. Give an intuitive reason why the function $g(x) = x^5 - 4x^2 + 1$ must take on the value 7 somewhere in the interval $[1, 2]$. Hint: Compute $g(1)$ and $g(2)$. Prove this fact by applying the IVT to the function $f(x) = g(x) - 7$ on the interval $[1, 2]$. What other values must the function g achieve on $[1, 2]$?

   b. Generalize the idea behind part a to complete, with understanding, the following statement.

   > *General Intermediate Value Theorem:*   Suppose the function g is _____ on the closed interval $[a, b]$.   For any real number d between _____ and _____ there exists a point c between _____ and _____ such that _____ .

11. Give a specific example of a function f that is not continuous on a closed interval $[a, b]$, but for which the conclusion of the General IVT on that interval $[a, b]$ still holds. Why does this not contradict the theorem?

Notes to Instructor   Lab 5: Investigating the Intermediate Value Theorem

Principal author: Ed Packel

Scheduling: This lab may be scheduled after the idea of a continuous function on an interval has been discussed.

Computer or calculator requirements: This lab requires the ability to plot graphs and zoom in to locate zeros of a function. A command for solving equations will expedite Problems 4 and 5.

Comments: The motivation behind this lab is that a surprising number of students seem, at exam time, to be confused about the idea, statement, and application of this first important theorem about continuous functions defined on a closed interval. It is hoped that by "discovering" the theorem, even with a "fill in the blank" format, the theorem will take on significantly more meaning to the student.

While this lab is more conceptual and less computational than others, it is still worthwhile to have the computer or calculator available for giving numerical values and graphs to support the theoretical conclusions. Perhaps the students should be alerted to the fact that thinking and understanding will be especially important in this lab.

Comments on implementing the lab: In Problem 2 the student will need to know how to define a "two-tiered" function. In Derive, version 2.0 or later, you can define the function in Problem 2a by

```
g(x) := if (x<=0, x^2+1, 1-x^2-x^4)
```

with attention restricted to the interval $[-3, 2]$. If you have only an older version of Derive, then you can build the function using characteristic functions. Thus the function can be defined by

```
g(x) := (x^2+1)*CHI(-3,x,0)+(1-x^2-x^4)*CHI(0,x,2)
```

(with some license for graphically irrelevant endpoint errors). In Mathematica, type

```
g[x_] := x^2+1/; And [-3<=x, x<=0]
g[x_] := 1-x^2-x^4/; And [0<x, x<=2]
```

with attention restricted to the interval $[-3, 2]$. In Maple, define this function by

```
g := proc(x) if x<=0 then x^2+1 else 1-x^2-x^4 fi end;
```

On the TI-81 or TI-85, define the function y1= (x^2+1)(x<=0) + (1-x^2-x^4)(x>0) on the interval $[-3, 2]$.

In Problem 5, the enterprising student may choose to draw the graph first, thereby simplifying the task of bracketing the zeros (and indicating that five zeros do appear to be present). Such foresight is praiseworthy, but it should be impressed upon the student that a careful proof that these are indeed zeros requires use of the IVT. Two of the zeros are at approximately -4.53674 and -1.81478. Full marks and glory to students who find these two and then observe that the function is odd, thereby determining the other three zeros.

Two of the standard corollaries of the IVT are given for the student to prove in Problems 8 and 9. This will be satisfying for those who see the point, though the instructor may need to downplay these two parts because of time or pedagogical constraints. In any event they should be formally developed in the first class following the lab.

Lab 6: Relationship between a Function and Its Derivative

Goal

- Given the graph of a function, to be able to visualize the graph of its derivative.

Before the Lab

In this laboratory, you will be asked to compare the graph of a function like the one in Figure 1 to that of its derivative. This exercise will develop your understanding of the geometric information that f' carries.

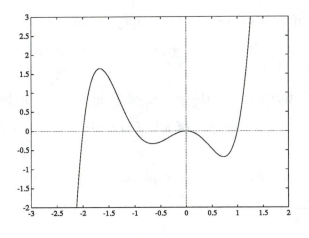

Figure 1: $f(x) = x^2(x-1)(x+1)(x+2)$

You will need to bring an example of such a function into the lab with you, one whose graph meets the x-axis at four or five places over the interval $[-3, 2]$. During the lab, your partner will be asked to look at the graph of your function and describe the shape of its derivative (and you will be asked to do the same for your partner's function). One way to make such a function is to write a polynomial in its factored form. For example, $f(x) = x^2(x+1)(x-1)(x+2)$ is the factored form of the function in Figure 1. Its zeros are at 0, 1, -1, and -2.

31

1. Give another example of a polynomial g of degree at least 5 with four or five real zeros between -3 and 2. You will use this polynomial in Problem 4.

   a. Your polynomial: $g(x) = $ _____.

   b. Its zeros: _____.

   c. Its derivative: $g'(x) = $ _____. If your function is complicated, you may want to use the computer to calculate its derivative.

In the Lab

2. Let $f(x) = x^2(x^2 - 1)(x + 2) = x^5 + 2x^4 - x^3 - 2x^2$.

   a. Find the derivative of f. Plot the graphs of both f and f' in the same viewing rectangle over the interval $[-2.5, 1.5]$.

   Answer the following questions by inspection of this graph:

   b. Over what intervals does the graph of f appear to be rising as you move from left to right?

   c. Over what intervals does the graph of f' appear to be above the x-axis?

   d. Over what intervals does the graph of f appear to be falling as you move from left to right?

   e. Over what intervals does the graph of f' appear to be below the x-axis?

   f. What are the x-coordinates of all of the high points and low points of the graph of f?

   g. For what values of x does the graph of f' appear to meet the x-axis?

3. Let $f(x) = \dfrac{x}{1 + x^2}$.

   a. Find the derivative of f. Plot the graphs of both f and f' in the same viewing rectangle over the interval $[-3, 3]$.

   b. Answer the same set of questions as in parts b-g above.

4. On the basis of your experience so far, write a statement that relates where a function is rising, is falling, and has a high point or low point to properties you have observed about the graph of its derivative.

5. Now let g be the function that your lab partner brought into the lab. (If you have no partner, just use your own function.) In this problem you will use your statement from Problem 4 to predict the shape of the graph of g', given only the shape of the graph of g.

   a. Have your lab partner produce a plot of the graph of g over the interval $[-3, 2]$. Your partner may need to adjust the height of the window to capture all of the action. On the basis of this plot, use your conjecture to imagine the shape of the graph of g'. In particular, find where g' is above, where g' is below, and where g' meets the x-axis. Carefully sketch a graph of both g and your version of g' on your data sheet, labeling each graph.

   b. Now have your lab partner plot the graph of g'. Add a sketch of the actual graph of g' onto your drawing. Compare your graph with the computer drawn graph. How did you do?

   c. Reverse roles with your lab partner and do parts a and b again.

6. Consider the function $f(x) = |x^2 - 4|$. A graph of the function will help you answer these questions.

   a. There are two values of x for which the derivative does not exist. What are these values, and why does the derivative not exist there?

   b. Find the derivative of f at those values of x where it exists. To do this, recall that f can be defined by $f(x) = \begin{cases} x^2 - 4, & |x| \geq 2, \\ 4 - x^2, & |x| < 2. \end{cases}$ You can compute the derivative for each part of the definition separately.

   c. Give a careful sketch of f and f' (disregarding the places where f' is not defined) over the interval $[-4, 4]$. Does your conjecture from Problem 4 still hold? Do you need to make any modifications?

Further Exploration

7. Consider the function $f(x) = 2^x$. Some people think that $f'(x) = x2^{x-1}$. On the basis of your conjecture explain why this cannot be true.

8. This laboratory has given you experience in using what you know about the shape of the graph of a function f to visualize the shape of its derivative function f'. What about going backwards? Suppose that your partner had given you the graph of f', would you be able to reconstruct the shape of the graph of f? If f' is positive, for example, does your conjecture enable you to rule out certain possibilities for the shape of f? The graph in Figure 2 is a sketch of the derivative of f. Use your conjecture to construct a possible graph for the function f itself. The important part of this problem is neither the actual shape that you come up with, nor its position in the xy-plane, but your reasons for choosing it. Why isn't there a unique function that has f' for its derivative?

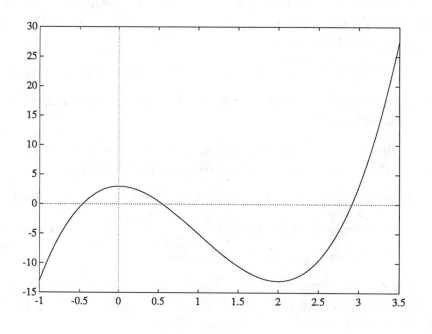

Figure 2: The graph of $y = f'(x)$

9. Figure 3 shows the graphs of three functions. One is the position of a car at time t minutes, one is the velocity of that car, and one is its acceleration. Identify which graph represents which function and explain your reasoning.

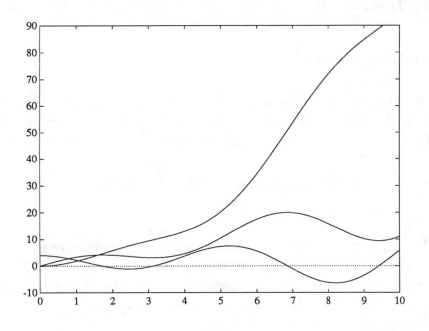

Figure 3: Position, velocity and acceleration graphs

Notes to Instructor   Lab 6: Relationship between a Function and Its Derivative

Principal author: John Fink

Scheduling: This lab should be scheduled before the students discuss the relationship between the sign of the derivative and the shape of the function. The student needs to be able to compute the derivatives, either by hand, calculator or computer.

Computer or calculator requirements: Ability to plot two functions in the same window.

Comments: The purpose of this lab is to give students experience in thinking about the relationship between the graphs of a function and its derivative. They are asked to go back and forth many times between the two graphs. The goal is that, by the end of the lab, they will be able to expect a certain shape for f', given enough detail about the shape of f.

In class, they should then be able to use their conclusion from this lab to understand the basic connections between increasing and decreasing functions and the sign of the derivative. These results should now appear geometrically obvious.

Lab 7: Linking Up with the Chain Rule

Goal

- To understand the Chain Rule for computing derivatives.

In the Lab

1. The Power Rule for computing derivatives says that $\dfrac{d}{dx}x^2 = 2x$. Thus we should not be surprised to learn that one factor of $\dfrac{d}{dx}\left(f(x)\right)^2$ is $2f(x)$. The purpose of this problem is to determine the other factor needed to yield the correct derivative.

   a. Functions of the form $g(x) = (ax + b)^2$ are simple enough that we can algebraically determine the other factor in the derivative. Expand $(ax + b)^2$ and differentiate the resulting expression term by term. Rewrite your answer to $g'(x)$ in the form $2(ax + b)k$. What is k? What part of the original expression can be differentiated to give k?

   b. For functions of the form $g(x) = (ax^2 + bx + c)^2$ we can use a computer or symbolic calculator to follow the same procedure of expanding and differentiating term by term to write the derivative as $g'(x) = 2(ax^2 + bx + c)k(x)$. What is $k(x)$? What part of the original expression can be differentiated to give $k(x)$?

   The next two problems use the method of zooming to approximate the derivative of a function f at a point a. To do this, use the computer to graph the function f on an interval containing $x = a$. Then zoom in around the point $\left(a, f(a)\right)$ until your graph resembles a straight line. Pick a point on the curve other than the point $\left(a, f(a)\right)$ and estimate its coordinates. Calculate the slope of the line through these two points. This will be a numerical approximation to $f'(a)$.

2. The derivative of $f(x) = \big(g(x)\big)^n$ contains the factor $n\big(g(x)\big)^{n-1}$. In this problem we will determine what the other factor might be.

   a. Complete the following table. In the first column you are given functions of the form $f(x) = \big(g(x)\big)^n$ where $g(x) = x^2 - 3x$. The evaluation of the second column can be done by hand. Use zooming to approximate the slope of f at $x = -1$ for the third column. In the last column enter your best guess of an integer or simple fraction you can multiply $n\big(g(-1)\big)^{n-1}$ by to get $f'(-1)$.

function $f(x) = \big(g(x)\big)^n$	value of $n\big(g(x)\big)^{n-1}$ at $x = -1$	approximation to $f'(-1)$	correction factor
$(x^2 - 3x)^2$			
$(x^2 - 3x)^3$			
$\sqrt{x^2 - 3x}$			

   b. In this problem the correction factor is the same for all three functions in part a. Where does this number come from?

   c. Consider the function $f(x) = (x^3 - 2x^2 - 2x + 6)^3$. Without using the computer, make a conjecture as to the value of $f'(1)$. Explain the reasoning behind your answer.

   d. In general $\dfrac{d}{dx}\big(g(x)\big)^n = n\big(g(x)\big)^{n-1}k(x)$. What is $k(x)$? What part of the original expression can be differentiated to give $k(x)$?

3. This problem examines the derivative of $f(x) = \sin\big(g(x)\big)$ for several choices of $g(x)$. Since $\dfrac{d}{dx}\sin x = \cos x$, we expect, and rightly so, that $\cos\big(g(x)\big)$ will be one factor of $\dfrac{d}{dx}\sin\big(g(x)\big)$. The question is: What else is needed?

   a. Complete the following table. In the first column you are given functions of the form $f(x) = \sin\big(g(x)\big)$. The evaluation in the second column can be done by calculator or computer. Use zooming to approximate the slope of f at $x = 3$ for the third column. In the fourth column enter your best guess of an integer or simple fraction you can multiply $\cos\big(g(3)\big)$ by to get $f'(3)$. The last column asks for the value of $g'(3)$. This can be computed exactly by hand.

function $f(x) = \sin(g(x))$	value of $\cos(g(x))$ at $x = 3$	approximation to $f'(3)$	correction factor	value of $g'(3)$
$\sin(2x)$				
$\sin(\frac{1}{2}x + 3)$				
$\sin(x^2)$				

b. Without using the computer, make a conjecture as to the form of the derivative of $f(x) = \sin(x^3)$ at $x = 3$? Explain your reasoning.

c. In general, $\dfrac{d}{dx}\sin(g(x)) = k(x)\cos(g(x))$ for some function $k(x)$. What do you think $k(x)$ equals? What part of the original expression can be differentiated to give $k(x)$?

Further Exploration

4. a. Write $(g(x))^2$ as $g(x)g(x)$ and apply the product rule to find the derivative of $(g(x))^2$.

b. Write $(g(x))^3$ as $(g(x))^2 g(x)$ and apply the product rule and your result in part a to find the derivative of $(g(x))^3$.

c. Let n be a positive integer. Suppose the pattern you observed in parts a and b continues to hold for the derivative of $(g(x))^{n-1}$. Write an expression for $\dfrac{d}{dx}(g(x))^{n-1}$. It should be in the form given in Problem 2d.

d. Now write $(g(x))^n$ as $(g(x))^{n-1}g(x)$. Use the product rule and the result of part c to find the derivative of $(g(x))^n$ in the form given in Problem 2d. By the Principle of Mathematical Induction, you are now justified in concluding that this pattern continues from one natural number to the next. Thus it holds for all natural numbers.

Notes to Instructor   Lab 7: Linking Up with the Chain Rule

Principal author: Anita Solow

Scheduling: After the Chain Rule is introduced in class.

Computer or calculator requirements: Ability to graph functions and zoom in on a point of interest. Problem 1*b* is easier with symbolic capability.

Comments: Students often use the Chain Rule more out of courtesy to the instructor than from any belief that it makes a difference. This lab attempts to convince students that the Chain Rule is needed when taking a derivative of a composition. It provides examples to demonstrate that omitting the second factor leads to answers that are algebraically or geometrically absurd.

Caution your students against using a large number of decimal places when they approximate the derivatives in Problems 2, 3 and 4.

Lab 8: Sensitivity Analysis

Goals

- To appreciate a classic textbook problem as a simplified version of a more complex commercial problem.

- To use the derivative as a rate of change to measure how sensitively one quantity depends upon another.

Before the Lab

1. Obtain a standard sized $(8\frac{1}{2}'' \times 11'')$ sheet of notebook paper. Lay it on a flat surface in front of you.

   a. Fold the bottom half up.

   b. Fold the left half over onto the right half.

   c. Fold the top left corner down along a 45° line from the top right corner to a point which should be $1\frac{1}{4}''$ from the lower left corner.

   d. Choose a point along the diagonal fold.

   e. Remove the top corner by a horizontal cut from the chosen point to the right edge of the folded paper.

   f. Unfold the paper. You should have congruent squares cut out of each corner of the sheet.

   g. Measure (in inches) the length x of the side of one of these squares.

   h. Fold up the four flaps and tape along the four cut edges to form a rectangular box with an open top.

   i. The height of the box is x. Measure (in inches) the other two dimensions of the box. How are they related to x?

   j. Compute the volume of the box (in cubic inches).

In the Lab

2. a. Tabulate the values of x and corresponding volumes for the boxes made by all the students in your lab.

    b. Plot a graph to show the relation between the values of x and the corresponding volumes. Estimate the value of x that seems to give the largest volume.

3. Find a formula for the volume of the box in terms of the variable x. What is the smallest allowable value for x in this situation? What is the largest? What volume results in each of these endpoint cases? Differentiate the volume function with respect to x. Find the critical points by setting the derivative equal to zero and solving for x. Check that one of your critical points gives a maximum volume. Call that critical point $x_{\max}$. Does the resulting volume agree with the estimate you made in Problem 2?

The general problem of optimizing one quantity under the constraint of fixed resources is present in nearly all aspects of life. The box problem is but a stylized sample of this type of problem. Although it is a classical example appearing in nearly every calculus textbook, it may nevertheless be of some importance to designers in the packaging industry. The remainder of this lab will direct you to examine some of the kinds of questions that typically arise in an optimization problem in an industrial or commercial setting.

4. Do you think anyone had a sheet of paper exactly $8\frac{1}{2}''$ wide? Certainly the dimensions of the original rectangle affect the critical point $x_{\max}$. Thus if you were negotiating a contract to manufacture containers with the volume as determined in Problem 3, you should request a certain tolerance in case your rectangles are slightly large or small.

    a. Solve the box problem again starting with a rectangle that is $11''$ long, but 2% wider than $8\frac{1}{2}''$. Notice that $x_{\max}$ is actually a function of the width w of the paper. Compare this value of $x_{\max}$ with the value in Problem 3 by computing the change $\Delta x_{\max}$ in the quantity $x_{\max}$. Compute the ratio of $\Delta x_{\max}$ to the change Δw in w as w increases 2% from $8\frac{1}{2}''$.

    b. Start with the formula for the volume of the box in terms of x and w. Use the symbolic capabilities of your computer to determine an expression for $x_{\max}$ in terms of w. Check that $\dfrac{dx_{\max}}{dw}$ evaluated at $w = 8.5$ is close to the difference quotient you evaluated in part a.

The derivative gives the rate of change of one quantity with respect to the change in another quantity. The size of the derivative is often more meaningful when compared to the sizes of the two quantities. In such cases we are led to consider the rate of change in the relative sizes of the two quantities. This is known as the *sensitivity* of one quantity to another.

5. In Problem 4a you computed $\Delta x_{\max}$, the change in $x_{\max}$ caused by a relative increase $\dfrac{\Delta w}{w}$ of 2% in w. Divide this change in $x_{\max}$ by the original value of $x_{\max}$ to obtain the relative change $\dfrac{\Delta x_{\max}}{x_{\max}}$ in $x_{\max}$. Compute the ratio of these two relative rates of change as an estimate of the sensitivity of $x_{\max}$ to w.

The *sensitivity* of one quantity to another is defined to be the limit of the ratio between the relative changes in the two quantities. We can easily relate the sensitivity to the derivative. In the box problem for example,

$$\lim_{\Delta w \to 0} \frac{\dfrac{\Delta x_{\max}}{x_{\max}}}{\dfrac{\Delta w}{w}} = \lim_{\Delta w \to 0} \frac{\Delta x_{\max}}{\Delta w} \frac{w}{x_{\max}} = \frac{dx_{\max}}{dw} \frac{w}{x_{\max}}.$$

6. In Problem 5 you used the ratio of the relative rates of change to estimate the sensitivity. In practice this is often turned around. The derivative used in the above formula gives the sensitivity. This then allows us to estimate the relative rate of change in one quantity as the sensitivity times the relative rate of change in another quantity. For example if the sensitivity is 3.7, then for small changes $\dfrac{\Delta x_{\max}}{x_{\max}}$ will be approximately 3.7 times $\dfrac{\Delta w}{w}$.

   a. For the box problem use the derivative formula to compute the sensitivity of $x_{\max}$ to w when the width is 8.5″. How does this compare with your estimate in Problem 5?

   b. Use your result from part a to estimate the value of $x_{\max}$ if w is 2% larger than 8.5″.

   c. Use your result from part a to estimate the percent increase in w that will give a 2% increase in $x_{\max}$.

7. a. Obtain a formula for the maximum volume V_{max} as a function of w. Notice that w appears directly in this formula; it also appears as part of x_{max}. Compute $\dfrac{dV_{max}}{dw}$. Compute the sensitivity of the maximum volume to the width of the rectangle when the width is 8.5″.

    b. If the width of the rectangle is 1% too small, estimate the percentage error in the maximum volume of the resulting box.

    c. A sensitivity of 1 or 2 or less indicates that a small change in the relative size of one quantity will produce a roughly comparable relative change in the dependent quantity. A sensitivity in the range of 5 to 10 or above indicates that a small relative change in one quantity will result in a relative change 5 to 10 or more times as great in the dependent quantity. Based on your results above, how concerned would you be about being able to produce a box with the promised volume?

Further Exploration

8. This problem concerns the sensitivity of both x_{max} and the maximum volume to the length of the rectangle.

    a. Compute the sensitivity of x_{max} to the length of the rectangle when the length is 11″.

    b. A 2% change in the length will lead to roughly what percent change in x_{max}?

    c. Compute the sensitivity of the maximum volume to the length of the rectangle when the length is 11″.

    d. If the length of the rectangle is 1% too small, estimate the percentage error in the maximum volume of the resulting box. How large a volume can you count on getting if you are likely to encounter error of up to 1% in the dimensions of the rectangle?

9. Suppose the manufacturing process is set up so that the box-cutting machine always makes a cut of size x_{max} as determined in Problem 3 for the standard $8\frac{1}{2}″ \times 11″$ sheet of paper. The height of the box will now be fixed, but its length and width will still vary according to the actual size of the paper being fed into the machine. With this dimension of the box fixed, what is the sensitivity of the volume to each of the dimensions of the paper? Explain in intuitive terms how these results differ from your results in Problems 7a and 8c.

Notes to Instructor   Lab 8: Sensitivity Analysis

Principal author: Robert Messer

Scheduling: This lab is intended to supplement the discussion of applied maximum/minimum problems. Students need to be able to set up and solve the standard textbook examples. They should also have a good working understanding of how derivatives are related to approximate rates of change. This lab is rather lengthy. Consider scheduling two lab sessions to give students time to think about the new concept of sensitivity. Alternatively, the first three or four problems can be done as a classroom demonstration.

Computer or calculator requirements: This lab makes essential use of symbolic differentiation, solving equations, and substitution of values into expressions. Only the most meticulous student would have any hope of correctly carrying out the necessary computations by hand. With a computer algebra system, however, students can concentrate on the concepts involved in sensitivity analysis without getting bogged down in the computational details.

Lab 9: Newton's Method

Goals

- To learn how to use Newton's Method to solve equations.

- To understand the geometry of Newton's Method.

- To see the importance of the initial guess.

Before the Lab

Very few equations $f(x) = 0$ can be solved exactly. You have learned methods and tricks for solving equations such as $x^2 - 6x + 6 = 0$, $x^4 - 5x^2 + 6 = 0$, and $\cos^2 x = \sin x$. (Spend a little time convincing yourself that you can solve these equations.) However, no general techniques exist for most equations, and we must settle for approximate solutions. The method we will study in this lab is attributed to Newton and uses the idea that the tangent line to a curve closely approximates the curve near the point of tangency.

Suppose we have a function f, and we want to solve $f(x) = 0$. To use Newton's Method, you must have an initial guess x_0. The next guess, x_1, is found at the intersection of the x-axis with the tangent line to $y = f(x)$ at $\left(x_0, f(x_0)\right)$. See Figure 1.

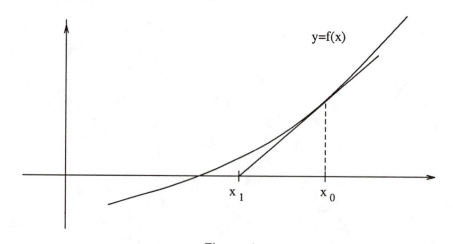

Figure 1

46

1. We need to find a formula for x_1.

    a. Use $\dfrac{\Delta y}{\Delta x}$ to find $f'(x_0)$, the slope of the tangent line to $y = f(x)$ at x_0, in terms of x_0, x_1, and $f(x_0)$.

    b. Solve for x_1 to get the first iteration of Newton's Method:

    $$x_1 = x_0 - \frac{f(x_0)}{f'(x_0)}.$$

    Once we have x_1, we repeat the process to get x_2 from x_1, x_3 from x_2, etc. If all goes well, the x_i's get closer and closer to the zero of f that we are seeking.

    c. Write a formula for x_2 in terms of x_1.

    d. Write a formula for x_{n+1} in terms of x_n.

    e. On Figure 2, sketch the appropriate tangent lines and show x_1, x_2, and x_3.

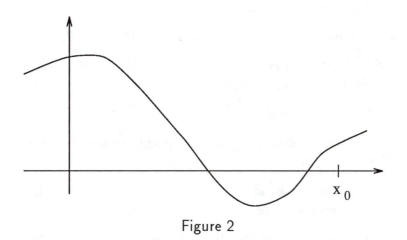

Figure 2

In the Lab

Computing Newton's Method by hand is tedious and error prone. Since there is a formula for each successive approximation for the solution of $f(x) = 0$, the procedure is easily performed by a computer or programmable calculator. In the lab you will follow your instructor's directions for using one of these tools to calculate solutions using Newton's Method.

2. The first application of Newton's method is to solve the equation $x^3 - 4x^2 - 1 = 0$.

    a. Graph the function $f(x) = x^3 - 4x^2 - 1$ and notice that the above equation has only one solution.

    b. Use Newton's Method with $x_0 = 5$. What solution did you find?

    c. Sketch, by hand, the first three iterations of Newton's Method on a graph of f.

3. You can use Newton's Method to find square roots of numbers. For example, to find $\sqrt{n}$, solve the equation $f(x) = 0$ where $f(x) = x^2 - n$.

    a. Find $\sqrt{15}$ using Newton's Method, specifying what value you used for x_0.

    b. Check your answer with your calculator or computer. Many calculators use Newton's Method, with an initial guess of 1, to take square roots.

    The main difficulty in using Newton's Method occurs in the choice of the initial guess, x_0. A poor choice can lead to a sequence $x_1, x_2, \ldots$ that does not get at all close to the solution you are seeking.

4. Go back to the equation $x^3 - 4x^2 - 1 = 0$ from Problem 2.

    a. Let $x_0 = 2$. What seems to be happening? Sketch the first three iterations of Newton's Method on a graph of $y = x^3 - 4x^2 - 1$.

    b. Let $x_0 = 0$. What happens?

5. In this problem we are going to find the point on the curve $y = 1/x$ that is closest to the point (1,0).

    a. Find the function (in the single variable x) that gives the distance from any point on the curve $y = 1/x$ to the point (1,0).

    b. Compute the derivative of the distance function that you found in part a.

    c. Display the necessary work to show that finding the critical points of the distance function leads to solving the equation $x^4 - x^3 - 1 = 0$.

    d. Now use Newton's Method to find all of the solutions of the equation $x^4 - x^3 - 1 = 0$.

    e. What point on the curve $y = 1/x$ is closest to the point (1,0)?

Further Exploration

6. Another method of approximating the zeros of a function f is to use the computer to graph the function $y = f(x)$, zoom in around a place where the graph hits the x-axis, and use the computer to digitize the coordinates of this point. Let's call this the *graphical* method of solving equations.

    a. Use the graphical method to solve the equation $x^3 - 4x^2 - 1 = 0$. (This is the same equation that you solved using Newton's Method in Problem 2.) Keep track of the answers that you get with successive zooming.

    b. Discuss the advantages and disadvantages that you think the graphical method has in relation to Newton's Method. Which would you rather use, and why?

7. This problem examines the sensitivity of Newton's Method to the choice of x_0. Let $f(x) = x^3 - x$. Clearly, the equation $x^3 - x = 0$ has the three solutions -1, 0, and 1.

    a. Argue that if $x_0 > 1/\sqrt{3}$, then Newton's Method will converge to the solution 1. Therefore, by symmetry, if $x_0 < -1/\sqrt{3}$, Newton's Method will converge to the solution -1. Please be sure to discuss what happens if $x_0 = \pm 1/\sqrt{3}$.

    b. Demonstrate algebraically that if we start Newton's Method with $x_0 = \pm 1/\sqrt{5}$, then $x_1 = \mp 1/\sqrt{5}$ and $x_2 = \pm 1/\sqrt{5}$. Therefore, if we start with $x_0 = \pm 1/\sqrt{5}$, we do not converge to a solution. In this case, $\pm 1/\sqrt{5}$ are called *period 2 points*.

    c. The interesting chaotic behavior occurs when $1/\sqrt{5} < x_0 < 1/\sqrt{3}$, or, by symmetry, when $-1/\sqrt{3} < x_0 < -1/\sqrt{5}$. Fill in the following table to convince yourself of the sensitivity of Newton's Method to the choice of x_0.

x_0	Solution found
0.577	
0.578	
0.460	
0.466	
0.44722	
0.44723	

Notes to Instructor   Lab 9: Newton's Method

Principal author: Anita Solow

Scheduling: Newton's Method can either be used as an application of derivatives or as an example of sequences of numbers.

Computer or calculator requirements: Graphing functions, built-in Newton's Method program (see below).

Comments: The purpose of this lab is to emphasize the geometry behind Newton's Method. Students should understand why the method is sensitive to the initial guess.

You may wish to check your students' work for Problem 5b before they attempt the following part. Otherwise, they may waste a good deal of time attempting to get their incorrect equation in the requested form.

The graphical method of solving equations (Problem 6) is rarely discussed, but it is certainly the most intuitively obvious approach when one has good quality computer graphics. The accuracy of the answer will depend largely on the quality of the graphics.

For more details about the chaotic behavior of Newton's Method that is briefly mentioned in Problem 7 in the Further Exploration section, see Philip Straffin, "Newton's Method and Fractal Patterns," in *Application of Calculus,* Volume 3 of *Resources for Calculus*, MAA Notes Series, (1993).

Since Newton's Method gives approximate solutions to equations, you will need to give your students an idea of accuracy. Four decimal places are more than ample to demonstrate the method. You might suggest a limit of 20 iterations when using Newton's Method.

Comments on implementing the lab: We strongly recommend that the instructor program Newton's Method before the lab so that it is available to the students. The following will work in Maple after a function f is defined.

```
newton := proc(x0,epsilon,n)
   local k, oldx, newx, fd;
   oldx := evalf(x0);
   fd := diff(f(x),x);
   for k from 1 to n do
      newx := evalf(oldx - f(oldx)/subs(x=oldx,fd));
```

```
            if abs(newx - oldx)<epsilon then RETURN(newx); fi;
            oldx := newx;
            print(k, newx);
        od;
    end:
```

In Mathematica, there is a built-in function to perform Newton's Method. Once you have defined the function $f[x]$, type

```
    FindRoot[{f[x] == 0}, {x, x0}]
```

where x0 is your initial guess. With this built-in command, you will not see each iterate.

In Derive, version 2.0 or higher, you can use the following command to perform Newton's Method.

```
    Newton(u,x,x0,n) := iterates(x - u/dif(u,x),x,x0,n)
```

where u is the expression for which a root is desired, x0 is the initial guess, and n is the number of iterations. If you have only an older version of Derive you can perform Newton's Method by the following set of commands. To make them as easy to follow as possible, I will give the instructions for performing Newton's Method to solve the equation $x^3 - 4x^2 - 1 = 0$. First `Author f(x) := x^3 - 4x^2 -1`. Then use `Calculus, Derivative` to take its derivative. With the derivative highlighted, `Author g(x) := F3 button`. At this point, $g(x)$ has been defined to be the derivative of $f(x)$, namely, $3x^2 - 8x$. Define a new function by

```
    N(x) := x - f(x)/g(x)
```

To calculate the answer from Newton's Method after 4 iterations, beginning with the initial guess of 5, either type `N(N(N(N(5))))` or evaluate `N(5)`, and then apply `N` to the answer three more times.

Lab 10: Indeterminate Limits and l'Hôpital's Rule

Goals

- To recognize limits of quotients that are indeterminate.

- To understand l'Hôpital's rule and its application.

- To appreciate why l'Hôpital's rule works.

Before the Lab

In this laboratory, we are interested in finding limits of quotients in cases which are referred to as *indeterminate*. This occurs, for example, when both the numerator and the denominator have limit 0 at the point in question. We refer to this kind of indeterminacy as the $\frac{0}{0}$ case.

One of the standard limit theorems allows us to compute, under favorable conditions, the limit of a quotient of functions as the quotient of the limits of the functions. More formally,

$$\lim_{x \to a} \frac{f(x)}{g(x)} = \frac{\lim_{x \to a} f(x)}{\lim_{x \to a} g(x)}$$

provided limits for f and g exist at a and $\lim_{x \to a} g(x) \neq 0$.

1. For three of the limits below, use the result stated above to evaluate the limits without the help of a calculator or computer. For each of the other three, explain why the limit theorem does not apply, say what you can about the limit of the quotient, and indicate which are indeterminate in the $\frac{0}{0}$ sense described above.

   a. $\displaystyle\lim_{x \to 2} \frac{x^2 + 1}{x - 4}$

   b. $\displaystyle\lim_{x \to 0} \frac{e^x - 1}{\sin x + 3}$

   c. $\displaystyle\lim_{x \to 2} \frac{\sqrt{x^2 + 5}}{\sin \pi x}$

   d. $\displaystyle\lim_{x \to 2} \frac{1 - \frac{2}{x}}{\sin \pi x}$

   e. $\displaystyle\lim_{x\to 3}\frac{x^2-9}{\sqrt{1+x}}$

   f. $\displaystyle\lim_{x\to a}\frac{f(x)-f(a)}{x-a}$, where f is any function that is differentiable at a

In the Lab

2. a. Plot the function $\dfrac{\ln x}{x^2-1}$ and, from the graph, determine or estimate the value of the indeterminate $\displaystyle\lim_{x\to 1}\frac{\ln x}{x^2-1}$.

    b. Repeat the procedure suggested in part a to obtain the limit for Problem 1d.

     If indeterminate quotients were always of such a specific nature and we were able and willing to use graphical or numerical estimates, there would be little need to pursue these matters further. Often, however, indeterminates occur in more abstract and general situations. Thus we seek a correspondingly general approach that will apply to a wide variety of indeterminate situations.

3. The result we will be exploring is known as *l'Hôpital's Rule*. One version of it says that if $\displaystyle\lim_{x\to a}\frac{f(x)}{g(x)}$ is indeterminate, but f and g have derivatives at a with $g'(a)\neq 0$, then $\displaystyle\lim_{x\to a}\frac{f(x)}{g(x)}=\frac{f'(a)}{g'(a)}$.

    a. By taking the appropriate derivatives, apply the above result to the quotient $\dfrac{\ln x}{x^2-1}$ at the point $a=1$. Also check the result graphically by plotting, on the same axes, the quotient of the functions f and g defined by $f(x)=\ln x$ and $g(x)=x^2-1$ along with the quotient of their derivatives (not the derivative of the quotient!) as called for by l'Hôpital's Rule. What should happen, according to l'Hôpital's Rule, in your graphs? Does it indeed happen?

    b. Do the same as above for the quotient of functions given in Problem 1d.

4. In this question we look at why the $f'(a)$ and $g'(a)$ arise in l'Hôpital's Rule. We restrict ourselves to the indeterminate case where both numerator and denominator approach 0 (*i.e.*, the $\frac{0}{0}$ case). Thus we assume that

    i. $\lim\limits_{x \to a} f(x) = 0$ and $\lim\limits_{x \to a} g(x) = 0$, and

    ii. f and g are both differentiable at a with $g'(a) \neq 0$.

a. Carefully explain why the assumptions made above ensure that $f(a) = 0$ and $g(a) = 0$.

b. Under the above assumptions, carefully justify each of the equalities in the line below.

$$\frac{f'(a)}{g'(a)} = \frac{\lim\limits_{x \to a} \dfrac{f(x) - f(a)}{x - a}}{\lim\limits_{x \to a} \dfrac{g(x) - g(a)}{x - a}} = \lim\limits_{x \to a} \frac{\dfrac{f(x) - f(a)}{x - a}}{\dfrac{g(x) - g(a)}{x - a}}.$$

c. Now build upon parts a and b to complete a proof for the $\frac{0}{0}$ case of l'Hôpital's Rule.

5. *Fun and games with l'Hôpital's Rule.*

a. Use l'Hôpital's rule to compute $\lim\limits_{x \to 0} \dfrac{1 - e^{3x}}{\sin x + x}$. If you can, use the `limit` command on your computer to check your answer.

b. Let f be a function that is differentiable at a. Perhaps the most famous indeterminate of all is $\lim\limits_{x \to a} \dfrac{f(x) - f(a)}{x - a}$. What does l'Hôpital's rule say in this situation? Are you surprised?

c. Make a conjecture about continuing the procedure called for in l'Hôpital's rule in situations where both $f'(a)$ and $g'(a)$ are also 0. Apply your conjecture to compute $\lim\limits_{x \to 0} \dfrac{x - \sin x}{1 - \cos x}$.

6. Consider the sector of a unit circle with angle x (in radians) as pictured in Figure 1. Let $f(x)$ be the area of the triangle ABC, while $g(x)$ is the area of the curved shape ABC.

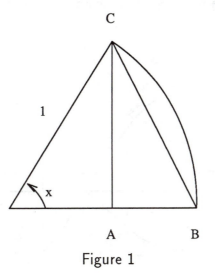

Figure 1

a. By thinking geometrically, try to make a guess about the limit of $f(x)/g(x)$ as the angle x approaches 0.

b. Show that $f(x) = \frac{1}{2}(\sin x - \sin x \cos x)$ and $g(x) = \frac{1}{2}(x - \sin x \cos x)$.

c. Using the result stated in part b, compute the actual limit you guessed at in part a.

Further Exploration

7. In Problem 4 you developed an analytic proof of l'Hôpital's rule. In this problem, you will develop a geometric justification that resembles l'Hôpital's original argument.

a. To start with an easier case, suppose f and g are straight lines that go through the point $(a, 0)$ with slopes m_1 and m_2, respectively. Therefore the equations for f and g are $f(x) = m_1(x - a)$ and $g(x) = m_2(x - a)$. Sketch a picture representing this situation. What is $\lim\limits_{x \to a} \dfrac{f(x)}{g(x)}$? (Do not use l'Hôpital's rule to evaluate this limit!)

b. Now assume that f and g are functions that satisfy the same conditions as those listed in the statement of Problem 4. Figure 2 shows the graphs of two such functions. If we zoom in near the point $(a, 0)$ we will see two straight lines. What are the slopes of those lines? Use these results along with the results from part a to complete a geometric justification for l'Hôpital's rule.

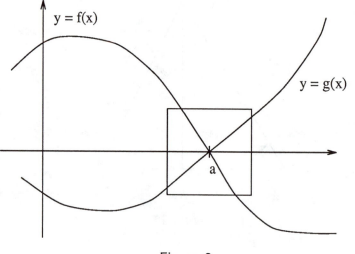

Figure 2

8. In checking l'Hôpital's rule graphically in Problem 3, you no doubt found that $\frac{f(x)}{g(x)}$ and $\frac{f'(x)}{g'(x)}$ approach the same value as x approaches a. This question asks you to investigate the *slopes* of these two quotient functions.

 a. By looking at the graphs in Problem 3 and other $\frac{0}{0}$ situations of your own choosing, try to discover a relationship between the above slopes.

 b. By using appropriate calculus techniques, prove your conjecture in part a. Hint: With $h(x) = \frac{f(x)}{g(x)}$ and $k(x) = \frac{f'(x)}{g'(x)}$, look carefully at and compare $h'(x)$ and $k'(x)$ as x approaches a.

Notes to Instructor   Lab 10: Indeterminate Limits and l'Hôpital's Rule

Principal author: Ed Packel

Scheduling: Any time after limits have been introduced and the definition of the derivative has been developed.

Computer or calculator requirements: Simultaneous plotting of two functions. Symbolic differentiation may be helpful at a few points, but is not required.

Comments: This laboratory calls for relatively little use of the computer. The emphasis is on understanding what one version of l'Hôpital's rule says and why it works.

Problem 6 is due to Gilbert Strang (M.I.T.). The author first saw the geometric justification of l'Hôpital's rule presented by Thomas Dick (Oregon State University).

The slope relationship called for in Problem 8 has the following formulation. If $\lim_{x \to a} f(x) = \lim_{x \to a} g(x) = 0$ and $g'(a) \neq 0$, then, with $h(x) \equiv f(x)/g(x)$ and $k(x) \equiv f'(x)/g'(x)$, it follows that $\lim_{x \to a} k'(x) = 2h'(a)$. It is worth noting that this relationship was actually "discovered" (undoubtedly not for the first time) in the process of developing this lab. Indeed, the authors observed that the relevant slopes seemed to have a regular relationship and were led to look for an analytic proof. The proof calls for careful use of the quotient rule, an application of l'Hôpital's rule, and wise use of the information given.

Lab 11: Riemann Sums and the Definite Integral

Goals

- To approximate the area under a curve by summing the areas of coordinate rectangles.

- To develop this idea of Riemann sums into a definition of the definite integral.

- To understand the relationship between the area under a curve and the definite integral.

Before the Lab

1. a. Figure 1 shows the graph of a function f on the interval $[a, b]$. We want to write an expression for the sum of the areas of the four rectangles that will depend only upon the function f and the interval endpoints a and b. The four subintervals that form the bases of the rectangles along the x-axis all have the same length; express it in terms of a and b. How many subinterval lengths is x_2 away from $a = x_0$? Write expressions for x_1, x_2, x_3, and x_4 in terms of a and b. What are the heights of the four rectangles? Multiply the heights by the lengths, add the four terms, and call the sum $R(4)$.

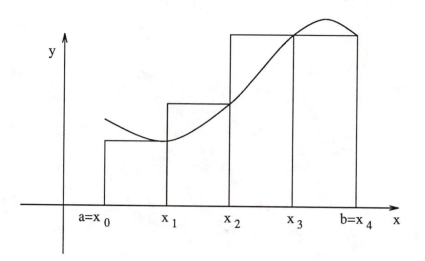

Figure 1

58

    b. Generalize your work in part *a* to obtain an expression for $R(n)$, the sum of rectangular areas when the interval $[a, b]$ is partitioned into n subintervals of equal length and the right-hand endpoint of each subinterval is used to determine the height of the rectangle above it. Write your expression for $R(n)$ using summation notation. In order to do this, first figure out a formula for x_k, the right-hand endpoint of the kth subinterval. Then check that your formula for x_k yields the value b when k takes on the value n.

    c. Let $L(n)$ denote the sum of rectangular areas when left-hand endpoints rather than right-hand endpoints are used to determine the heights of the rectangles. Add some details to Figure 1 to illustrate the areas being summed for $L(4)$. What modification in the expression for $R(n)$ do you need to make to get a formula for $L(n)$?

    d. Have your instructor check your formulas for $R(n)$ and $L(n)$.

In the Lab

2. Consider the function $f(x) = 3x$ on the interval $[1, 5]$.

    a. Apply your formulas for $R(4)$ and $L(4)$ to this function. In your notebook sketch a picture to illustrate this situation. Use geometry to determine the exact area A between the graph of f and the x-axis from $x = 1$ to $x = 5$.

    b. Now use the computer to get values for $R(n)$ and $L(n)$ for $n = 40$. How are the sizes of $R(40)$ and $L(40)$, related to the area A? Explain this relationship and make a conjecture about what should happen for larger and larger values of n. Test your conjecture with a few more values of n of your own choosing and record your results in a table.

3. Consider the function $f(x) = \frac{1}{\sqrt{x}}$ on the interval $[.1, 10]$. Plot the graph of f and use the ideas developed above to approximate the area A under the graph of f and above the given interval. What is the relation among $R(n)$, $L(n)$, and the area A now? Explain any difference that you see between the situation here and the situation in Problem 2.

4. Now let $f(x) = 4 - x^2$ on the interval $[-2, 2]$. Again plot the graph and estimate the area under it on this interval. How are the values of $R(n)$ and $L(n)$ related this time? What has changed and why? Does this prevent you from getting a good estimate of the area? Explain.

5. So far we have worked with functions that are nonnegative on their domains. There is, however, nothing in the formulas you have developed that depends on this. Here we consider what happens geometrically when the function takes on negative values.

   a. Consider the function $f(x) = 3x$ on the interval $[-4, 2]$. In your notebook sketch the graph of f on this interval and include appropriate rectangles for computing $R(3)$ and $L(3)$. On subintervals where the function is negative, how are the areas of the rectangles combined in obtaining the overall value for $R(3)$ or $L(3)$? What value do $R(n)$ and $L(n)$ seem to approach as n increases? How can you compute this value geometrically from your sketch?

   b. Now consider the function $f(x) = 3x^2 - 2x - 14$ on the interval $[2, 3]$. Use your computer to plot it. Determine the value that $R(n)$ and $L(n)$ seem to approach. Explain with the help of your graph why you think this is happening.

Here is a summary of the two important points so far:

- Approximation by rectangles gives a way to find the area under the graph of a function when that function is nonnegative over the given interval.

- When the function takes on negative values, what is approximated is not an actual area under a curve, though it can be interpreted as sums and differences of such areas.

In either case the quantity that is approximated is of major importance in calculus and in mathematics. We call it the *definite integral of f over the interval* $[a, b]$. We denote it by $\int_a^b f(x)\,dx$.

For well-behaved functions it turns out that we can use left-hand endpoints, right-hand endpoints, or any other points in the subintervals to get the heights of the approximating rectangles. Any such sum of areas of approximating rectangles (over any partition of $[a, b]$ into subintervals, equal in length or not) is called a *Riemann sum*. We obtain the definite integral as a limit of the Riemann sums as the maximum subinterval length shrinks to 0. In particular, for sums based on right-hand endpoints and equal length subintervals, we have $\lim_{n \to \infty} R(n) = \int_a^b f(x)\,dx$.

Similarly $\lim_{n \to \infty} L(n) = \int_a^b f(x)\,dx$.

6. For each of the definite integrals below, use either left or right-hand Riemann sums to approximate its value. Make your own decision about what values of n to use. Then use the definite integral command on your computer algebra system to obtain another approximation. Comment on the degree to which the values agree.

   a. $\displaystyle\int_0^1 e^x \, dx$

   b. $\displaystyle\int_1^3 (x^3 - 3x^2 - 2x + 3) \, dx$

   c. $\displaystyle\int_{-1}^3 \sin(x^2) \, dx$

Further Exploration

7. So far we have formed Riemann sums using left-hand or right-hand endpoints of the subintervals. Another Riemann sum uses rectangles whose heights are obtained by taking the midpoint of each subinterval. Develop a summation formula for this method and use it on the integrals in Problem 6 with the same values of n you used there. Which of the two methods, midpoint or endpoint Riemann sums, seems to give better approximations to the definite integral? Try to give a general explanation for your conclusion.

8. Figure 2 suggests a way of computing the volume of a unit sphere by adding up volumes of cylindrical disks as x ranges from -1 to 1.

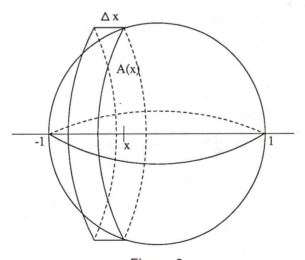

Figure 2

a. Let $A(x)$ denote the area of the cross-sectional circle at the point x. Write a formula for $A(x)$.

b. Argue that $R(n) = \sum_{k=1}^{n} A(x_k)\Delta x$ approximates the volume of the sphere, where the x_k are equally spaced and divide the interval into n subintervals each of length Δx.

c. Use the formulas developed in the lab to compute $R(10)$, $R(20)$, and $R(40)$.

d. Since $R(n)$ is a Riemann sum, $\lim_{n \to \infty} R(n)$ should give the exact value for the volume of a sphere of radius 1. Write the definite integral and obtain its value using the definite integral command. Does it agree with the usual formula for the volume of a sphere?

Notes to Instructor   Lab 11: Riemann Sums and the Definite Integral

Principal author: Ed Packel

Scheduling: This lab can be scheduled just prior to the in-class introduction of the definite integral.

Computer or calculator requirements: Graph plotting, summation, approximate values of definite integrals.

Comments: This lab motivates and develops the definite integral as a limit of Riemann sums. The approach is geometric and invites the student to find the appropriate extension of area for functions taking on negative values. The final section of the lab discusses the definite integral, introduces the notation $\int_a^b f(x)\,dx$, and asks students to try the definite integral command on their computer. This should pave the way for a more careful and complete treatment of these ideas in class.

Comments on implementing the lab: Students should work on developing the formula for $R(n)$ on their own (ideally the night before the lab). However, they should be given strategies for getting the computer algebra system to compute $R(n)$. Here is a sequence of commands to do this in Derive. Note that the third line establishes f as a function.

```
d(n):=(b-a)/n
x(k,n):= a + k*d(n)
f(x):=
r(n):= d(n)*sum(f(x(k,n)),k,1,n)
```

Having entered this, one need only declare values for a and b, and assign the appropriate function rule to f(x). Then r(40), for instance, will be the Riemann sum. A more direct approach in Derive, if your students are comfortable with more variables and like to take large bites, is:

```
f(x):=
r(a,b,n):=(b-a)/n*sum(f(a+k*(b-a)/n),k,1,n)
```

In Maple the following built-in procedure will work.

```
with(student);
rightsum(f(x), x=a..b, n);
value(");
evalf(");
```

Similarly, in place of `rightsum` you can use `leftsum`. The rightsum command forms the sum but does not evaluate it. The last two commands do that.

Here are Mathematica commands to compute R(n) after `f[x_]` is defined.

```
r[a_,b_,n_] := (b-a)/n * Sum[f[a+k*(b-a)/n], {k, 1, n}]
```

In Problem 3 and beyond there is an intentional vagueness about what values of n to use. It seems appropriate to let the student take charge of this. After testing the lab on a computer algebra system, the instructor may elect to provide additional guidance to avoid outrageous requests and to allow easier checking and comparison of results.

Lab 12: Area Functions

Goals

- To introduce the idea of area under a curve as a function.
- To extend this idea to accumulation functions.
- To make a conjecture about the derivatives of such functions.

Before the Lab

In this lab we will start with a positive function, use its graph as the upper boundary of a region in the plane, and create a new function in terms of the area of the region. The x-coordinate of a vertical line that serves as the right-hand boundary of the region will be the variable in this new function.

1. Sketch the line with slope 3 that passes through the origin. For any positive value of x, consider the triangle in the first quadrant bounded above by the line, below by the x-axis, and on the right by the vertical line x units from the y-axis. What is the area of this triangle when $x = 4$? What is the area when $x = 17$? Find a general formula for the area in terms of x.

2. Sketch the line passing through $(0,9)$ and $(9,0)$. For any value of x between 2 and 9, consider the trapezoid bounded above by the line and below by the interval $[2, x]$. The parallel sides of the trapezoid will be the vertical lines passing through the horizontal axis at $(2,0)$ and $(x,0)$.

   a. Find a formula for the area of the trapezoid in terms of x.

   b. Confirm that your formula gives the value $-\frac{15}{2}$ when $x = 1$. Interpret this negative number in relation to the area of the trapezoid above the interval $[1, 2]$ on the x-axis.

   c. Interpret your formula for $x = 2$.

   d. Confirm that your formula gives the value 20 when $x = 12$. Check that this is smaller than the value of your formula at $x = 9$. How much smaller? Interpret your result in terms of areas. What is happening when $x > 9$?

65

3. In the first two problems you have developed formulas for what we will call the *accumulation function*. This function extends the idea of the area function to situations where the original function is negative and to the case where the ending value is to the left of the starting value along the x-axis. The concept of accumulation is the counterpart to the concept of instantaneous change you have studied as the derivative. Differentiate the accumulation functions you found in Problems 1 and 2. Do you recognize the derivatives?

In the Lab

In Problems 1 and 2, you were able to use geometry to find the accumulation functions. In general, if you are given a function f, you will not necessarily be able to find a formula for the accumulation function F. If f is positive, the accumulation F over an interval $[a, x]$ represents the area under the graph of f and above the interval $[a, x]$. We can use the notation of the definite integral to write $F(x) = \int_a^x f(t)\,dt$. The variable t is used inside the integral so as not to be confused with the x in the interval $[a, x]$. Your instructor will give you a method for graphing the accumulation function F. (You will never need to use a formula for F to do the remainder of this lab. You will use your computer to graph a numerical approximation to F.)

4. In Problem 1 you found the accumulation function for $f(x) = 3x$. Use your instructor's method for graphing accumulation functions and check that it agrees with the function you found geometrically.

5. Consider the semicircle given by the function $f(x) = \sqrt{1 - x^2}$. The graph is the upper half of a circle of radius 1 centered at the origin.

   a. Plot both the semicircle and the accumulation function for f over the domain $[-1, 1]$. Use these graphs to do the remainder of this problem.

   b. Evaluate the accumulation function at $x = 0$ and $x = 1$ and compare your results with the exact area of a quarter circle and a half circle of unit radius.

   c. Where does the accumulation function appear to have horizontal tangent lines? Where does the accumulation function increase the fastest? Relate these properties of the accumulation function to properties of the original function $f(x) = \sqrt{1 - x^2}$.

6. Consider the function $f(x) = \sin(x^2 - 3x + 1)$. The graph of this function exhibits some interesting behavior. Furthermore, mathematicians have proved that its accumulation function cannot be written in terms of the standard functions of calculus. Thus we must rely on computations of approximate values to find out what it is.

   a. Plot a graph of $f(x) = \sin(x^2 - 3x + 1)$ over the domain $[0, 4]$. Also plot the accumulation function over this domain.

   b. Make a table for the accumulation function showing the intervals where it is increasing, the intervals where it is decreasing, and its maximum and minimum points. In another column of this table describe where f is positive, negative, and zero.

   c. Explain any relations you observe. Which relations appear to be coincidences and which do you think will hold between any function and its accumulation function?

7. Look back over your work for this lab and make a conjecture about the derivative of the accumulation function. Cite evidence for your conjecture.

Further Exploration

8. Try using the geometric definition of the accumulation function to determine its derivative. Start by sketching a typical positive, continuous function f. Let F denote the accumulation function, which we can interpret as the area under the graph of f above an interval $[a, x]$.

   a. Because we intend to establish a basic derivative formula for a new function F, we will need to go back to the definition of $F'(x)$ as the limit of a difference quotient. Write the formula for $F'(x)$ using the definition of the derivative.

   b. Interpret the numerator as the difference between the areas of two overlapping regions. Sketch an example and shade in the vertical strip whose area corresponds to the numerator. Interpret the denominator as the width of the vertical strip. What is the geometric significance of the quotient?

   c. How does the limit of this quotient compare with the value of f at x? State your conclusion as a general formula for $F'(x)$. Congratulations! You have just given an intuitive proof of the Fundamental Theorem of Calculus, which some people think is the most important intellectual achievement of recorded history.

Notes to Instructor   Lab 12: Area Functions

Principal author: Robert Messer

Scheduling: Students should have a good grasp of the relation between a function and its derivative. In particular they should be able to match the graph of a function with the graph of its derivative. This lab should be scheduled to precede any classroom discussion of the Fundamental Theorem of Calculus.

Computer requirements: This lab requires the ability to graph two functions on the same set of axes. The computer should also be able to graph an approximation to the definite integral of a function over intervals with varying endpoints. There are several methods of doing this, and your choice will depend on the speed of your system. One method is to have the computer plot a numerical approximation to $\int_a^x f(t)\,dt$ for each value of x in the domain using the built-in numerical integration facilities. If this method is too slow, the midpoint rule with fewer than a dozen subintervals will do. The use of other numerical integration rules will be postponed until Lab 18. Instructions for the midpoint rule follow.

In Derive, first `Author` an arbitrary function `f` and then `Author` a midpoint rule `S(a,b,n)` for `f` over the interval $[a,b]$ based on `n` subintervals.

```
f(x) :=
S(a,b,n) := (b-a)/n sum(f(a + (k-.5)(b-a)/n),k,1,n)
```

The students can then supply the actual definition of the function `f` and approximate the accumulation function by `S(0,x,8)` for example.

In Mathematica define the function `f` and then the midpoint rule

```
S[a_,b_,n_]:=((b-a)/n)*Sum[f[a + (k-.5)(b-a)/n],{k,1,n}]
```

In Maple define the function `f`. To use the midpoint rule, type

```
mid :=  proc(a,b,n)
   local k, h;
   h := evalf((b-a)/n*sum(f(a + (k-.5)*(b-a)/n),k=1..n));
end:
```

Use `mid(0,x,8)` to get an approximation for the accumulation function over the interval $[0,x]$.

Lab 13: Average Value of a Function

Goals

- To develop the notion of the average value of a continuous function over a given interval.

- To gain some geometric insight into the location of the average within the range of values taken on by a continuous function.

Before the Lab

What do we mean when we say, "The average temperature today was sixty degrees"? We clearly intend the single number 60 to represent the entire range of temperatures for the day. How is this number to be computed?

Taking the average of a finite sample of temperatures is easy to do if there aren't very many: just add up the temperatures and divide by the sample size, twenty-four, for instance, if the samples were taken hourly. But it is not so obvious how to proceed if we want to take into account the continuous variation in temperatures that actually occurred that day.

In this lab we will use the definite integral in one of its most basic and powerful roles: as a generalization of addition to infinite sets of numbers.

A *thermograph* is a continuous record of the temperature throughout the day. The graph in Figure 1 is the thermograph for August 29, 1990, taken from a device on the roof of the Des Moines Register building.

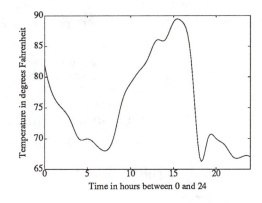

Figure 1: Thermograph

The newspaper summarizes this graph by giving the hourly temperature readings (Table 1) and what it calls the "daily average."

Des Moines Weather--August 29, 1990

1 a.m. 77	7 a.m. 68	1 p.m. 86	7 p.m. 70
2 a.m. 75	8 a.m. 70	2 p.m. 86	8 p.m. 70
3 a.m. 73	9 a.m. 76	3 p.m. 89	9 p.m. 69
4 a.m. 70	10 a.m. 79	4 p.m. 89	10 p.m. 67
5 a.m. 70	11 a.m. 81	5 p.m. 83	11 p.m. 67
6 a.m. 69	noon 83	6 p.m. 67	midnight.....67

Temperatures: High 89 at 4 p.m.; low 67 at 6 p.m.; average 78.

Table 1

1. The newspaper's version of the average should really be called the "mid-range" temperature. It is the temperature halfway between the high and low for the day. A closer approximation to the number we seek might be obtained by taking the average of the twenty-four hourly recorded temperatures in Table 1.

   a. Compute this average. Which of the numbers, the mid-range or the average of the hourly temperatures, seems to be a better measure of the overall impact of the day's temperatures?

   b. Draw a graph similar to the one in Figure 1 which has the same mid-range temperature, but which you would judge to represent a distinctly hotter day. Why does your graph seem to belong to a hotter day than the graph in Figure 1?

We could get an even closer approximation to the average temperature for the day by taking the average of 48 temperatures, recorded at half-hour intervals throughout the day:

$$\frac{\sum_{i=1}^{48} T_i}{48}$$ where T_i is the temperature taken in the ith half hour.

The general strategy is clear. Divide the day into n segments, each of length $\frac{24}{n}$ hours, record the temperature at a particular time in each segment, and take the average of this finite set of numbers in the standard way: $\frac{\sum_{i=1}^{n} T(x_i)}{n}$, where $T(x)$ is the temperature at time x and x_i is the time at which the temperature is recorded in the ith segment of the day.

The limit of these averages will be the number we seek. Its connection with the integral is made by means of the Riemann sum as follows:

$$\text{Average} \;=\; \lim_{n \to \infty} \frac{\sum_{i=1}^{n} T(x_i)}{n} \;=\; \frac{1}{24} \lim_{n \to \infty} \sum_{i=1}^{n} T(x_i)\,\frac{24}{n} \;=\; \frac{1}{24} \int_{0}^{24} T(x)dx.$$

Any continuous function f over the interval $[a, b]$ has an average value that can be computed in just this way:

$$\textit{Average value of } f \textit{ over } [a, b] = \frac{1}{b - a} \int_{a}^{b} f(x)dx.$$

The one thing that might appear mysterious to you about this formula is the $\dfrac{1}{b - a}$ that appears in front of the integral. Make sure you know why it is there, and how it got there.

Any continuous function f over the interval $[a, b]$ also has a mid-range value (why?) that is computed by taking the value halfway between the extreme values of the function over $[a, b]$.

$$\textit{Mid-range value of } f \textit{ over } [a, b] = \frac{f(x)_{\max} + f(x)_{\min}}{2}.$$

In this lab you will be asked to estimate the average value of functions like the one shown in Figure 2.

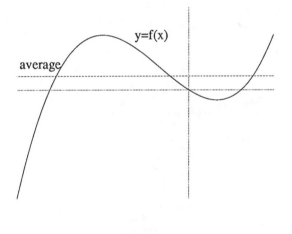

Figure 2

You will need to bring an example of such a function into the lab with you, one that changes its concavity at least once in the chosen interval. During the lab your partner will be asked to look at the graph of your function and to estimate its average value over the interval you have chosen (and you will be asked to do the same for your partner's function). One way to make such a function is to add a linear function to a cubic polynomial written in its factored form.

2. a. Your function $g(x)$: _____.

    b. The interval over which you plan to take the average of $g(x)$: _____.

In the Lab

3. *A geometric interpretation.* Let $f(x) = x$ over the interval $[0, 2]$.

    a. Use the definition to compute the average value of f over the interval $[0, 2]$. Is it what you expect it to be? How does it compare with the mid-range value of f?

    b. Draw a horizontal line at the level of the average value on your graph of f over the interval $[0, 2]$. You should get a picture like the one in Figure 3. Without using any calculus, show that the area of region A is the same as the area of region B.

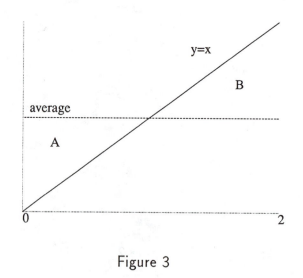

Figure 3

4. Consider $f(x) = x^2$ over the interval $[0, 2]$.

 a. Use calculus to compute the average value of f over the interval $[0, 2]$. How does it compare with the mid-range of f over this interval?

 b. Draw a horizontal line at the level of the average value on your graph of f over the interval $[0, 2]$. You should get a picture like the one in Figure 4. Use calculus to show that the area of region A is the same as the area of region B. One method to do this is to find the area of A and the area of B and then to show that they are equal. A simpler method is to show that $\int_0^2 (f(x) - \text{average value of } f) \, dx = 0$. Why does this simpler method work?

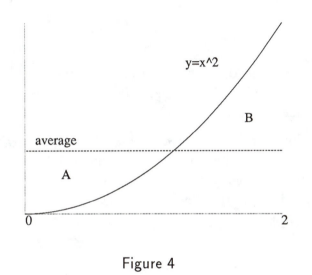

y=x^2

B

average

A

0   2

Figure 4

5. Let $f(x) = 4x - x^2$ over $[0, 4]$.

 a. Compute the mid-range and average values. How do they compare?

 b. Draw a horizontal line at the level of the average value on the graph of f to get a picture like the one in Figure 5. Does the area of region B equal the combined areas of regions A and C? Show it by using the simpler method explained in Problem 4b.

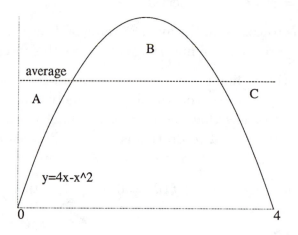

Figure 5

6. *Eyeballing the average.* By now you probably have enough experience with the average value to be able to estimate it visually. Test your eyes on the function that your partner brought into the lab.

   a. Have your lab partner produce a plot of the graph of g over the interval chosen. Your partner may need to adjust the height of the window to capture all of the action. When you are satisfied with the picture on the screen, copy it carefully into your notebook. Then sketch a horizontal line at the level where you think the average is. Use only the graph of the function, not its formula.

   b. Now have your partner compute the actual value of the average, and plot it along with the graph of g. How did you do?

   c. Reverse roles with your lab partner and do parts *a* and *b* again.

   d. In your write-up for this problem, explain how you estimated the average value of your partner's function.

Further Exploration

7. *Definitions — old and new.* New definitions in mathematics are always suspect until they have been shown to be consistent with their earlier use. The purpose of this problem is to test this new definition of "average value" in the context of rectilinear motion.

    Here is a typical situation involving distance and velocity.

    > "A stone is thrown upward from a 54 foot high platform with an initial velocity of 30 feet per second. After t seconds have elapsed, its velocity is $v(t) = 30 - 32t$. It lands on the ground 3 seconds after it was thrown."

    a. Use the new definition to compute the average velocity of the stone over its trajectory.

    b. Most textbooks define *average velocity* to be "the change in position divided by the change in time." Compute the average velocity of the stone again, using this definition. Your two answers for the average velocity should agree.

    c. Use the Fundamental Theorem of Calculus to explain the equality of your answers in parts a and b.

8. *Average vs. mid-range.* In the problems in this lab you noticed that the mid-range value was likely to be different than the average value of the function. In this question we will explore whether there is anything about the shape of the graphs of the functions that could account for this difference.

    a. The graph of the function $f(x) = e^x$ is concave up on $[0, 1]$. Is the mid-range above or below the average?

    b. The graph of the function $f(x) = \sin x$ is concave down over the interval $[0, \pi]$. What is the relationship between the mid-range and the average value over this interval?

    c. State clearly (in the form of a conjecture) the connections that you think hold between concavity, average value, and mid-range value for a continuous function over the interval $[a, b]$.

    d. Prove this conjecture. The trapezoid rule will help you get started.

Notes to Instructor   Lab 13: Average Value of a Function

Principal author: John Fink

Scheduling: Any time after the Fundamental Theorem of Calculus has been discussed. Students also need to know how to set up an integral for finding the area between two curves.

Computer or calculator requirements: Plotting two functions on the same graph. Numerical computation of definite integrals could be used, but is not necessary.

Comments: The purpose of this lab is to give students a chance to see the integral in one of its most basic applications as a means of measuring the average behavior of a function over an interval. This lab differs from most of the others of this collection in its use of the machine. Nearly all of the problems can be done without a computer or calculator. Nevertheless, it will be handy to have one available to calculate the averages which students are asked to estimate in Problem 6.

Lab 14: Arc Length

Goals

- To develop the idea of approximating arc length by the sum of the lengths of straight line approximations to the curve.

- To compute arc length using an integral.

Before the Lab

The idea of the arc length of a curve is very easy to understand. You have had experience calculating the length of straight lines and circles. Intuitively, the length of the curve on the left in Figure 1 should be the length of the straight line we get on the right if we take the ends of the curve and pull it taut. Although this idea is easy to understand, it is not much good as a practical method of computation.

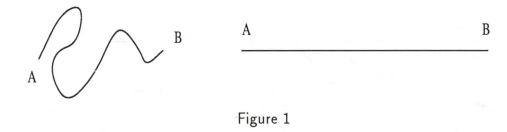

Figure 1

1. We start by reviewing the method for calculating the length of a straight line segment.

   a. Find the length of the straight line segment from $(1, 2)$ to $(5, 4)$, shown in Figure 2. The lengths of the two sides of the right triangle are $4 - 2 = 2$ and $5 - 1 = 4$. You can use the Pythagorean Theorem to determine the length of the hypotenuse.

   b. Now generalize this example to find a formula for the length of the segment of the straight line $y = mx + c$ from $x = a$ to $x = b$. This is the line segment connecting the points $(a, \underline{\hspace{1cm}})$ and $(b, \underline{\hspace{1cm}})$. Show the necessary work to get the length equal to $(b - a)\sqrt{1 + m^2}$. Check your answer to part a using this formula.

   c. Write an expression for the length of the curve in Figure 3 made up of four straight line segments with slopes m_1, m_2, m_3 and m_4.

77

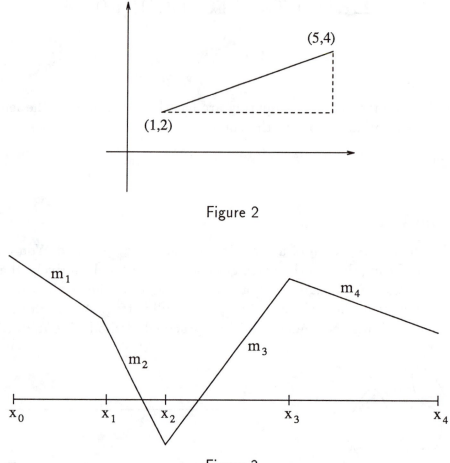

Figure 2

Figure 3

2. We are now ready to try to find the length of a curve defined by $y = f(x)$ for $a \le x \le b$, where the curve is not necessarily a straight line. Consider, for example, the graph $y = \sin x$ for $0 \le x \le \pi$, shown in Figure 4.

   a. A crude approximation to the length of the curve would be the straight line distance between the endpoints of the curve. Compute this distance. Is this value bigger or smaller than the actual arc length of this graph?

   b. A better approximation would be found by dividing $[0, \pi]$ into two pieces and adding the straight line distance from $(0,0)$ to $(\frac{\pi}{2}, 1)$ to the distance from $(\frac{\pi}{2}, 1)$ to $(\pi, 0)$. See Figure 4. Compute this distance. How is it related to the answer in part a? How is it related to the actual arc length?

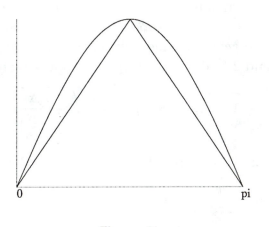

Figure 4

c. To get a more accurate answer, we will subdivide $[0, \pi]$ more finely yet. Copy the graph of $y = \sin x$ from Figure 4 into your notebook and sketch the straight line approximation to the curve using four subintervals of equal size. On another copy of this curve, sketch the straight line approximation to the curve using eight subintervals of equal size. In the lab, you will use the computer to calculate the lengths of these approximations.

In the Lab

3. Use the computer commands provided by your instructor for calculating straight line approximations to the arc length. For each of the following four functions, determine the length of the straight line approximations to arc length using 2, 4, 8, 25, 50, and (if your computer is fast enough) 100 subintervals of equal size. Make your best guess as to the length of the curve, accurate to at least two decimal places. Record your data in a suitably labeled table.

   a. $f(x) = \sin x$ for $0 \le x \le \pi$

   b. $f(x) = \sqrt{9 - x^2}$ for $0 \le x \le 3$

   c. $f(x) = \dfrac{x^3}{6} + \dfrac{1}{2x}$ for $1 \le x \le 3$

   d. $f(x) = e^x$ for $0 \le x \le 1$

   e. You can find the exact answer for the arc length of one of these curves without using calculus or the computer. Which curve is it, and what is the exact answer? (Hint: One is the equation of a well-known geometric object.)

It is possible to write an integral for the length of a curve $y = f(x)$ for $a \le x \le b$. This is done carefully in your textbook using the Mean Value Theorem. The derivation relies on the geometric ideas we have been using in this lab. We have seen in Problems 1 and 2 that the arc length is approximated by

$$\sum_{k=1}^{n} \sqrt{1 + m_k^2} \left(\frac{b-a}{n} \right) = \sum_{k=1}^{n} \sqrt{1 + \left(\frac{\Delta y_k}{\Delta x} \right)^2} \, \Delta x$$

where $\Delta x = \dfrac{b-a}{n}$ and the slope of the kth straight line segment is $\dfrac{\Delta y_k}{\Delta x}$. As $\Delta x \to 0$, we know that $\dfrac{\Delta y_k}{\Delta x} \to f'(x)$. Therefore the arc length equals

$$\lim_{\Delta x \to 0} \sum_{k=1}^{n} \sqrt{1 + \left(\frac{\Delta y_k}{\Delta x} \right)^2} \, \Delta x = \int_{a}^{b} \sqrt{1 + f'(x)^2} \, dx.$$

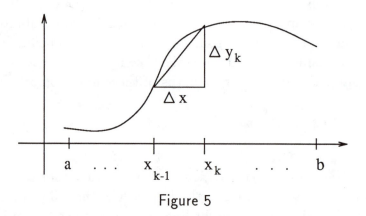

Figure 5

For example, $\displaystyle\int_{0}^{\pi} \sqrt{1 + \cos^2 x} \, dx$ gives the length of the curve $f(x) = \sin x$ for $0 \le x \le \pi$. Convince yourself that this is the correct integral.

Although we now have an integral formula for arc length, in practice it is difficult (or impossible) to apply the Fundamental Theorem of Calculus to most of the resulting integrals. This leaves us needing a numerical integration technique to approximate the value. For example, we cannot integrate $\displaystyle\int_{0}^{\pi} \sqrt{1 + \cos^2 x} \, dx$ exactly, but we can use a computer to find that the arc length of $y = \sin x$ for $0 \le x \le \pi$ is approximately 3.8202.

4. a. Write the appropriate integrals for the arc lengths for the curves given in Problems 3*b*, 3*c*, and 3*d*.

   b. The integral for Problem 3*c* can be computed by hand. Do it.

   c. Use your computer to approximate the value of the arc length integrals for the functions given in Problems 3*b* and 3*d*.

   d. Compare your answers to those you obtained in Problem 3.

Further Exploration

5. An approach to finding the arc length that people sometimes think of is the staircase method. This method is to partition $[a, b]$ into n pieces and approximate the arc length by the sum of the vertical and horizontal lines. See Figure 6. Unfortunately, this method does not work. Why not? What happens to the picture as n becomes arbitrarily large? What happens to the sum as n increases to infinity?

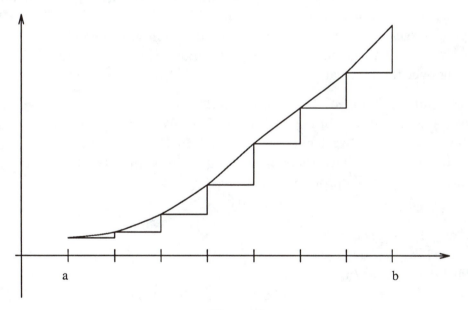

Figure 6

Notes to Instructor   Lab 14: Arc Length

Principal author: Anita Solow

Computer or calculator requirements: Computing and approximating definite integrals, built-in function for computing the sums of the lengths of piecewise linear approximations to a curve (see below).

Scheduling: This is an application of integration that is easy to understand and can be scheduled any time after the theory of the definite integral has been presented.

Comments: Of all the applications of integrals usually done, arc length is perhaps the easiest for students to understand geometrically. Unfortunately, the integrands that emerge from arc length rarely have recognizable antiderivatives. However, we can get good numerical answers using a computer.

Comments on implementing the lab: The integral for the arc length of the quarter circle of radius 3 (Problem 4c) is an improper integral. The accuracy of the answer will depend largely on the sophistication of your computer or calculator. If there are problems, students should be able to get a modestly accurate approximation by integrating over the interval $[0, 2.999]$, instead of $[0, 3]$.

To compute the sum of the lengths of the piecewise linear approximation to the curve, a computer algebra system is needed. We strongly recommend that the instructor program an arc length approximation function into the computer for students to use. We believe there is nothing to be gained (and much to be lost) having students type the program themselves.

Maple:

```
arclength := proc(a,b,n)
   local delx, m, s, k;
   delx := (b-a)/n;
   m := (f(a + k*delx) - f(a + (k-1)*delx))/delx;
   s := evalf(delx * sum(sqrt(1 + m^2), k=1..n));
end;
```

Derive: First `Author` an arbitrary function `f` and then `Author` a function to calculate the approximate arc length.

```
f(x) :=
Arc(a,b,n) := (b-a)/n * sum(sqrt(1 + ((f(a + k*(b-a)/n) -
   f(a + (k-1)*(b-a)/n))/((b-a)/n))^2),k,1,n)
```

Mathematica: Define the function f, and then estimate the arc length by

```
arclength[f_,a_,b_,n_] :=
    (
    delx = (b-a)/n;
    x[k_] := a + k*delx;
    m[k_] := (f[x[k]] - f[x[k-1]])/delx;
    length = N[delx*Sum[Sqrt[1 + m[k]^2], {k,1,n}]]
    )
```

Lab 15: A Mystery Function

Goals

- To investigate the effects of shrinking and stretching graphs.

- To use the idea of defining a function as the area under a curve.

- To develop some basic properties of a mystery function and use them to identify the function.

In the Lab

1. Recall how the period of the sine and cosine functions changes when the independent variable is scaled by a constant factor: if x is replaced by $\frac{x}{b}$, the graph is stretched horizontally by a factor of b changing the period from 2π to $2\pi b$. Go ahead and have the computer or calculator plot $y = \sin x$ and $y = \sin \frac{x}{3}$, for example, if you have any doubts about this.

   a. This stretching phenomenon does not depend on any special properties of the trigonometric functions or even on the fact that these functions are periodic. It actually works for any function. For example, use your computer or calculator to graph $g(x) = |x^3 - 6x^2 + 9x - 3|$ on the interval $[0, 4]$. Use the graph to approximate the x and y coordinates of all of the relative maxima and minima of this function on the interval $[0, 4]$. Now graph $y = g(\frac{x}{2})$ over the interval $[0, 8]$. What are the approximate x and y coordinates of all of the relative maxima and minima of this function on the interval $[0, 8]$? Repeat with the graph of $y = g(\frac{x}{.75})$ over the interval $[0, 3]$. Describe what happens to corresponding relative maximum and minimum points when x is replaced by $\frac{x}{b}$.

   b. Consider the regions bounded by the graph of $g(x) = |x^3 - 6x^2 + 9x - 3|$, the x-axis and the line $x = 4$. Use the graphs to determine how the total area of these regions compares with the total area of the regions bounded by the graph of $y = g(\frac{x}{2})$, the x-axis and the line $x = 8$. How do you think the total area of these regions compares with the total area of the regions bounded by the graph of $y = g(\frac{x}{.75})$, the x-axis and the line $x = 3$? Make a general statement about what you think happens to the area of any region when it is stretched horizontally by a factor of b.

c. In a similar fashion, we can plot $y = \frac{1}{c}g(x)$ to shrink the graph of any function $y = g(x)$ vertically by a factor of c. Try this out on your computer with the function $g(x) = |x^3 - 6x^2 + 9x - 3|$ over the interval $[0, 4]$. Compare its graph with the graphs of $y = \frac{1}{2}g(x)$ and $y = \frac{1}{5}g(x)$ over the same interval. How do the x-coordinates and y-coordinates of the critical points of g compare with those of $\frac{1}{2}g$ and $\frac{1}{5}g$? How do the areas of the regions change as you change from g to $\frac{1}{c}g$?

We want to apply your observations about stretching and shrinking to the function $f(x) = \frac{1}{x}$. This function has the special property that when we stretch its graph horizontally by a factor of b and then shrink it vertically by the same factor, we return to the original graph. Indeed, $\frac{1}{b}f(\frac{x}{b}) = \frac{1}{b}(\frac{1}{\frac{x}{b}}) = \frac{1}{b}(\frac{b}{x}) = \frac{1}{x}$. We will define a mystery function M in terms of the area of a region whose upper boundary is the graph of $y = \frac{1}{x}$. Your job is to exploit the special property of this graph to prove some of the basic properties of this new function, thus identifying it as a function you have undoubtedly encountered in your precalculus course.

2. a. Define $M(x)$ to be the area of the region in the first quadrant under the graph of $f(x) = \frac{1}{x}$ and above the interval $[1, x]$. Use vertical lines through the endpoints of the interval to bound the region on the left and right. In your notebook sketch the regions whose areas correspond to $M(2)$, $M(3)$, and $M(6)$ to be sure you understand how the right endpoint of the interval $[1, x]$ serves as the independent variable in this new function. What is the domain of M? Using integral notation we can write $M(2) = \int_1^2 \frac{1}{x}\,dx$, $M(3) = \int_1^3 \frac{1}{x}\,dx$, etc.

b. Your instructor will provide you with a method for approximating M. Use this method to estimate $M(x)$ for $x = .5, 1, 1.5, 2, 2.5, 3, 4, 4.5, 7.5, 12$. Feel free to trade some of these values with your neighbors in the lab. Record the results in a table. Can you explain why the minus sign appears in the value of $M(.5)$? Is your value of $M(1)$ reasonable in light of the geometric definition of M?

c. Add your estimates of $M(1.5)$ and $M(2)$. What value on your table is your sum close to? Check for other sums that match other values. What relation is suggested by your observations? State this relationship as clearly as you can.

3. a. In your notebook sketch the graph of $f(x) = \frac{1}{x}$ and shade in the region whose area defines $M(a)$ for a typical value of a. Stretch the picture horizontally by a factor of b. As you discovered in Problem 1a, the resulting curve is the graph of $y = f(\frac{x}{b}) = \frac{b}{x}$. Make a second sketch to illustrate how the original shaded region stretches to the region below the new curve and above the interval $[b, ab]$ on the x-axis. How does the area of this region compare to the area of the original region?

   b. On a third sketch, shrink the graph of $f(\frac{x}{b})$ vertically by a factor of b. As you discovered in Problem 1c, the resulting curve is the graph of $\frac{1}{b}f(\frac{x}{b})$. As demonstrated in the paragraph before Problem 2, this simplifies to the original function $f(x) = \frac{1}{x}$. Shade in the region between the new curve and the interval $[b, ab]$ and explain why its area is equal to $M(a)$.

   c. On this third sketch, use a different color to shade in the region above the interval $[1, b]$ whose area represents $M(b)$. The union of the two shaded regions in this sketch has an area equal to the function M evaluated at what single number? State your conclusion as an identity involving $M(a)$, $M(b)$, and $M(ab)$.

4. Let us use the geometric definition of M to compute its derivative. Because we intend to establish a basic derivative formula for a new function, we will need to go back to the definition of the derivative as the limit of a difference quotient.

   a. Write the formula for $\frac{d}{dx}M(x)$ in terms of the definition of the derivative.

   b. Interpret the numerator as the difference between the areas of two regions, one contained within the other. Sketch a typical situation and shade in the vertical strip whose area represents the difference in the areas of these regions. Interpret the denominator as the width of the vertical strip. What is the geometric significance of the quotient?

   c. How does the limit of this quotient compare with the graph of $f(x) = \frac{1}{x}$? State your conclusion as a general formula for $\frac{d}{dx}M(x)$.

   d. Using the same method you used previously to approximate the function M, compute the difference quotients $\frac{M(2.1)-M(2)}{2.1-2}$ and $\frac{M(2.01)-M(2)}{2.01-2}$ as approximations to $M'(2)$. How do these values compare with the derivative formula you obtained in part c? Check some difference quotients that will approximate $M'(x)$ for some other values of x.

5. The function we are calling M is a function that you have seen before in your precalculus course. State what function you think M is by identifying a familiar function that satisfies the property established in Problem 3c or whose derivative satisfies the formula you discovered in Problem 4c. If you are stuck, use either a precalculus or calculus book to help you.

Further Exploration

6. Think of a as the product $\frac{a}{b} \cdot b$. Apply the identity you derived in Problem 3c to this product to derive an identity for $M(\frac{a}{b})$.

7. Another identity that M satisfies is $M(a^r) = r\,M(a)$. Of course, we need to assume $a > 0$ in order for M to be defined. Here is an outline for proving this relation.

    a. Check that the formula holds for the trivial cases $r = 0$ and $r = 1$. Use the multiplication formula for M to prove the new identity for $r = 2$, $r = 3$, and $r = 4$. Argue that once you have the formula for any whole number r, you can extend it to hold for the next whole number $r + 1$. Conclude that the formula holds for any positive integer r.

    b. Use your results so far to verify that $M(a^{-n}) = -n\,M(a)$ if n is a positive integer. Conclude that $M(a^r) = r\,M(a)$ for any integer r.

    c. If r is a rational number, we can write $r = \frac{p}{q}$ where p and q are integers. Write a as $a^{1/q}$ to the power q, and use the previous result to conclude that $M(a^{1/q}) = \frac{1}{q}M(a)$. Now continue the string of equalities $M(a^{p/q}) = M((a^p)^{\frac{1}{q}}) = \cdots$ ending with $\frac{p}{q}M(a)$. Be sure you can justify each step based on results from earlier steps in this lab.

    d. If you have a clear idea of what a^r means when r is irrational, use your definition of a^r to extend the identity to all real numbers r. Otherwise, don't worry if you have not seen a definition of a^r for irrational exponents. You are well on your way to bridging this gap.

Notes to Instructor   Lab 15: A Mystery Function

Principal author: Robert Messer

Scheduling: This lab is best scheduled any time after the students have seen both the definition of the derivative and the integral.

Computer or calculator requirements: This lab requires the ability to graph several functions on the same set of axes. A method of numerical integration is needed.

Comments: The goal of this lab is for the student to identify the mystery function M as the natural logarithm by recognizing that they have the same properties.

Comments on implementing the lab: A computer or calculator should be used to approximate the integral for evaluating M. Either use your computer algebra system's built-in definite integration capability, or use a simple numerical method such as the midpoint rule with twenty or so subintervals. Other methods of numerical integration will be postponed until Lab 18. Instructions for the midpoint rule can be found in *Notes to Insructor* for Lab 12.

Lab 16: Exploring Exponentials

Goals

- To investigate the derivative of exponential functions.

- To define an exponential function as a limit.

- To define an exponential function as a series.

In the Lab

1. *Derivatives of exponential functions.* Exponential functions are functions of the form $f(x) = a^x$. They play a key role in the applications of mathematics. In a calculus course, we naturally wonder how to find $f'(x)$. Experiments with something as simple looking as $f(x) = 2^x$ quickly show that first guesses are generally wrong.

    a. Demonstrate the absurdity of thinking that $g(x) = x2^{x-1}$ might be the derivative of $f(x) = 2^x$. For example, have your computer plot the graphs of these two functions. Think about what the graph of 2^x would look like if its derivative were negative. Also look at $x = 0$ where $x2^{x-1}$ changes sign.

    b. We need to go back to the definition of the derivative to derive a formula for $f'(x)$ where $f(x) = 2^x$. Supply reasons for each of the following three steps:

    $$\frac{d}{dx}2^x = \lim_{h \to 0} \frac{2^{x+h} - 2^x}{h} = \lim_{h \to 0} \frac{2^x(2^h - 1)}{h} = 2^x \lim_{h \to 0} \frac{2^h - 1}{h}.$$

    Use your computer to gather evidence that $\lim_{h \to 0} \dfrac{2^h - 1}{h}$ exists. You might evaluate the quotient for small values of h or zoom in on the graph of the quotient near the y-axis. Notice that the derivative of f is simply this constant times the function itself.

    c. Modify the process in part b to find a formula for the derivative of $f(x) = 3^x$. Notice again that the derivative is a multiple of f, but with a different constant factor.

    d. Show in general with any positive number a as the base of an exponential function $f(x) = a^x$, that

    $$f'(x) = \frac{d}{dx}a^x = a^x \lim_{h \to 0} \frac{a^h - 1}{h}.$$

e.   We are interested in locating a value of a for which the multiplier $\lim\limits_{h \to 0} \dfrac{a^h - 1}{h}$ is equal to 1. Such a base will give an exponential function whose derivative is exactly itself. Check back to parts b and c to see that the base $a = 2$ gives a multiplier less than 1, while $a = 3$ gives a multiplier greater than 1. Try to narrow in on a value of a between 2 and 3 that gives an exponential function whose derivative is itself.

The number you have discovered in Problem 1e is commonly designated by e. This symbol was first introduced by the eighteenth century Swiss mathematician Leonhard Euler. With this notation, your discovery in Problem 1e is that if $f(x) = e^x$, then $f'(x) = e^x$.

2.   *An exponential function in the world of banking.*

a.   If you deposit money in a savings account paying interest at an annual rate of r, your deposit will grow by a factor of $1 + r$ after one year. (If the interest rate is 5%, then we use $r = .05$.) If, however, the bank compounds the interest in the middle of the year, your deposit will grow by a factor of $1 + \frac{r}{2}$ after the first six months, and another factor of $1 + \frac{r}{2}$ after the second six months. Thus the total growth is $(1 + \frac{r}{2})^2$. Verify that $(1 + \frac{r}{2})^2$ is slightly more than $1 + r$ for all nonzero values of r.

b.   By what factor will your deposit grow in one year if the bank compounds quarterly? If it compounds monthly? Daily? Hourly? The limiting value of the yearly growth factor as the number of compounding periods increases to infinity is $\lim\limits_{n \to \infty} (1 + \frac{r}{n})^n$. This is the growth factor used if the bank compounds continuously.

c.   Substituting $r = 1$ into the limit in part b gives $\lim\limits_{n \to \infty} (1 + \frac{1}{n})^n$. Use your computer or calculator to approximate the value of this limit. Where have you seen this number before?

d.   In part b we introduced the limit $\lim\limits_{n \to \infty} (1 + \frac{r}{n})^n$, which is a function of r. You evaluated this function for $r = 1$ in part c. If $r = 2$, we have $\lim\limits_{n \to \infty} (1 + \frac{2}{n})^n$. Use your computer or calculator to approximate the value of this limit. Compare your answer with the number e^2. Repeat for $r = 5$, comparing your answer with the number e^5. In fact, the function e^r can be defined by this limit for all values of r, namely, $e^r = \lim\limits_{n \to \infty} (1 + \frac{r}{n})^n$.

3. *An exponential function as a series.* Consider the infinite series

$$1 + \frac{x}{1!} + \frac{x^2}{2!} + \frac{x^3}{3!} + \frac{x^4}{4!} + \frac{x^5}{5!} + \cdots .$$

The exclamation mark in the denominators indicates the factorial of a positive integer. For example, $5! = 5 \cdot 4 \cdot 3 \cdot 2 \cdot 1 = 120$.

a. Substitute $x = 1$ into this series and sum the first 8 terms. What number do you obtain?

b. Repeat part a substituting $x = 2$ into the series. Where have you seen this number before in this lab?

c. Compute the derivative of the series (term by term) and show that it equals the series itself. How does this result relate to the result of Problem 1e? What function of x do you think this series represents? Give as much evidence as you can for your answer.

Further Exploration

4. For what value of a are the graphs of $y = a^x$ and $y = \log_a x$ tangent to one another? Recall that $\log_a x = \frac{\ln x}{\ln a}$. You should use your computer or graphing calculator to approximate the solution before you attempt to solve for a analytically.

Notes to Instructor   Lab 16: Exploring Exponentials

Principal author: Robert Messer

Scheduling: Any time after the definition of the derivative has been developed.

Computer or calculator requirements: Ability to graph functions.

Comments: This lab considers three independent approaches to the exponential function. The Further Exploration Problem requires knowledge of inverse functions. It was suggested by Richard Lane (University of Montana) and first appeared as Problem 814 in *Mathematics Magazine*, **44** (November-December 1971).

Lab 17: Patterns of Integrals

Goals

- To become familiar with the types of functions that are the antiderivatives of given families of functions.

- To get practice drawing conclusions from patterns of answers.

In the Lab

In this lab you will use a computer to calculate the antiderivative of several functions. Each problem concerns a family of related functions. You are to develop an intuition about the form of the antiderivative of each of these families.

1. a. Determine antiderivatives of the following rational functions.

        i. $\displaystyle\int \frac{1}{(x+1)(x+2)}\, dx$

       ii. $\displaystyle\int \frac{1}{(x+2)(x-7)}\, dx$

      iii. $\displaystyle\int \frac{1}{x(x-2)}\, dx$

      iv. $\displaystyle\int \frac{1}{(x-3)^2}\, dx$

     b. Based on the pattern of your answers to parts *i–iii*, determine an antiderivative for $\dfrac{1}{(x+p)(x+q)}$ where $p \neq q$. What changes do you need to make in your answer in the case where $p = q$?

2. a. Determine antiderivatives of the following powers of x times the exponential function.

       i. $\displaystyle\int x\, e^x\, dx$

       ii. $\displaystyle\int x^2 e^x\, dx$

      iii. $\displaystyle\int x^5 e^x\, dx$

iv. $\displaystyle\int x^7 e^x \, dx$

b. From your answers in part a, determine $\displaystyle\int x^n e^x \, dx$.

3. a. Determine antiderivatives of the following powers of x times the logarithm function.

i. $\displaystyle\int \ln x \, dx$

ii. $\displaystyle\int x \ln x \, dx$

iii. $\displaystyle\int x^3 \ln x \, dx$

iv. $\displaystyle\int x^6 \ln x \, dx$

b. Based on your answers in part a, determine an antiderivative of $x^n \ln x$.

4. a. Determine antiderivatives of the following powers of the cosine function.

i. $\displaystyle\int \cos x \, dx$

ii. $\displaystyle\int \cos^3 x \, dx$

iii. $\displaystyle\int \cos^5 x \, dx$

iv. $\displaystyle\int \cos^2 x \, dx$

b. Based on your answers to parts *i–iii*, determine the form of the antiderivative of $\cos^n x$, where n is a positive odd integer. (When giving your answer, do not worry about the value of the coefficients.) Note that the pattern is different when n is an even integer.

5. a. Determine antiderivatives of the following products of sine functions.

i. $\displaystyle\int \sin x \sin 2x \, dx$

ii. $\displaystyle\int \sin 2x \sin 5x \, dx$

iii. $\displaystyle\int \sin 6x \sin 2x \, dx$

b. Using your answers to part a, determine the form of $\int \sin ax \sin bx\, dx$.

Further Exploration

6. Refer to your work for Problem 1. Use partial fractions to prove that your formula for $\int \dfrac{1}{(x+p)(x+q)}\, dx$ is correct in the case where $p \neq q$.

7. Refer to your work for Problem 3.

   a. Does your formula for $\int x^n \ln x\, dx$ continue to work for negative values of n? Be careful about the case $n = -1$. If necessary, modify your formula.

   b. Use integration by parts to prove that your formula for $\int x^n \ln x\, dx$ is correct.

Notes to Instructor   Lab 17: Patterns of Integrals

Principal authors: Bonnie Gold and Anita Solow

Scheduling: This lab should be done any time after the definition of the indefinite integral is developed. It could come before or during a chapter on the techniques of integration or replace part of it. Students need some knowledge of exponential and logarithm functions. Partial fractions is needed for the first problem in Further Exploration and integration by parts is needed for Problem 7*b*.

Computer requirements: A computer algebra system that can calculate antiderivatives.

Comments: This lab was written with the philosophy that since computers can perform symbolic integration, there is little reason to teach our students the full range of integration techniques. For example, the ACM-GLCA syllabus suggests teaching only the methods of substitution and integration by parts.

The first question in this lab deals with integrating rational functions. The instructor could demonstrate in a later class where the computer's answers came from.

The major activity in this lab is pattern recognition. This is a skill that is heavily used by mathematicians but rarely encountered by the students in their courses. The idea for this lab came from Don Small and John Hosack's *Explorations in Calculus with a Computer Algebra System*, Mc-Graw Hill, Inc., 1991.

Lab 18: Numerical Integration

Goals

- To understand the geometry behind two methods of numerical integration, the Trapezoid Rule and Simpson's Rule.

- To gain a feel for the relative speeds of convergence of Riemann sums, Trapezoid Rule, and Simpson's Rule.

Before the Lab

Read through this lab and answer Problems 1, 2a, 3abc, and 4.

In the Lab

Suppose we want to calculate $\int_a^b f(x)\,dx$. The Fundamental Theorem of Calculus states that $\int_a^b f(x)\,dx = F(b) - F(a)$, where F is an antiderivative of f. For example, $\int_0^{\pi/2} \cos x\,dx = \sin(\pi/2) - \sin(0) = 1$. However, integrals such as $\int_0^{\pi/2} \cos\sqrt{x}\,dx$ and $\int_0^1 \sqrt{1 + 9x^4}\,dx$ still give us problems since we cannot find usable expressions for the antiderivatives of the integrands. In this lab we will explore several methods for computing numerical approximations to such integrals.

Riemann Sums

One approach that you have seen in the definition of an integral is to form a Riemann sum. In this method, we replace the area under the curve $y = f(x)$, $a \leq x \leq b$, by the area of some rectangles. In Figure 1 we have a picture of a Riemann sum using four subintervals of equal length, with the height of each rectangle being the value of the function at the left-hand endpoint of that subinterval.

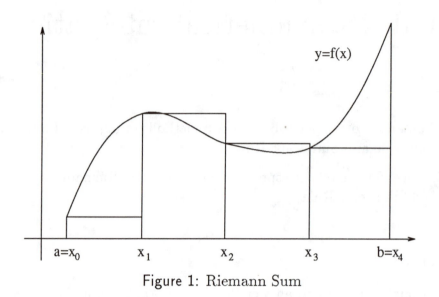

Figure 1: Riemann Sum

1. In this lab, we will be comparing several numerical answers for the value of $\int_0^1 5x^4 - 3x^2 + 1 \, dx$ with the exact answer obtained by direct integration. What is the exact answer to $\int_0^1 5x^4 - 3x^2 + 1 \, dx$?

2. Make a table to hold your answers for Problems 2, 4, and 5. Your table should include the method used, the number of subintervals, the approximation to the integral, the error in the approximation, and the width of a subinterval.

   a. Use a Riemann sum with $n = 4$ subintervals (by hand or calculator, showing all work) to approximate $\int_0^1 5x^4 - 3x^2 + 1 \, dx$. Be sure to specify whether you followed a left-hand or right-hand rule.

   b. Use a Riemann sum program on your computer to approximate the same integral with $n = 16$ subintervals.

   c. Find a value for n so that the Riemann sum gives an answer that is accurate to 0.001.

Trapezoid Rule

In Riemann Sums, we replace the area under a curve by the area of rectangles. However, the corners of the rectangles tend to stick out. Another method is to form trapezoids instead of rectangles.

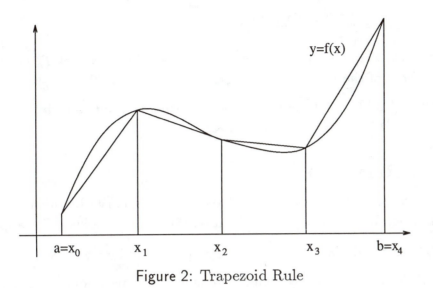

Figure 2: Trapezoid Rule

We will now develop the formula for the sum of the area of these trapezoids. This formula is known as the Trapezoid Rule.

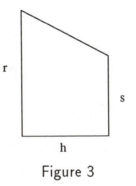

Figure 3

3. a. The formula for the area of a trapezoid is easy to derive. Divide the trapezoid in Figure 3 into a rectangle and a right triangle. The area of the rectangle is _____. The area of the triangle is _____. Show the algebra necessary to get the total area to be $\frac{h}{2}(r+s)$.

    b. Apply this formula four times to the four trapezoids in Figure 2. Let T_4 denote the sum of the areas of the four trapezoids. Show the algebra necessary to get that

$$T_4 = \frac{\Delta x}{2}\Big[f(x_0) + 2f(x_1) + 2f(x_2) + 2f(x_3) + f(x_4)\Big],$$

where $\Delta x = \dfrac{b-a}{4} = \dfrac{x_4 - x_0}{4}$.

c. If we use n equally spaced subintervals instead of 4, we let T_n be the sum of the areas of the n trapezoids. Derive a formula for T_n.

d. Repeat Problem 2 using the Trapezoid Rule, putting your data in the table.

Simpson's Rule

In the Trapezoid Rule, we replaced pieces of the curve by straight lines. In Simpson's Rule, we replace pieces of the curve by parabolas. To approximate $\displaystyle\int_a^b f(x)\,dx$, we divide $[a, b]$ into n equally spaced subintervals, where n is even. Simpson's Rule relies on the fact that there is a unique parabola through any three points on a curve. A picture of Simpson's Rule where $n = 4$ is given in Figure 4. The dashed line is the parabola through $\big(x_0, f(x_0)\big)$, $\big(x_1, f(x_1)\big)$ and $\big(x_2, f(x_2)\big)$, while the dotted line is the parabola through $\big(x_2, f(x_2)\big)$, $\big(x_3, f(x_3)\big)$ and $\big(x_4, f(x_4)\big)$.

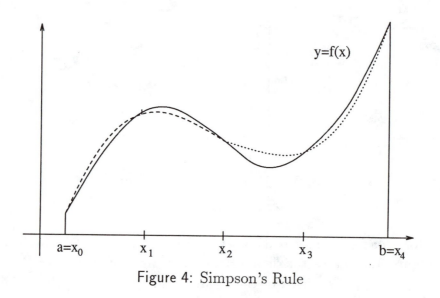

Figure 4: Simpson's Rule

The details are messy, but the area under the parabola through $\big(x_0, f(x_0)\big)$, $\big(x_1, f(x_1)\big)$ and $\big(x_2, f(x_2)\big)$ can be shown to be $\dfrac{\Delta x}{3}\big[f(x_0) + 4f(x_1) + f(x_2)\big]$, where $\Delta x = \dfrac{b-a}{4} = \dfrac{x_4 - x_0}{4}$.

4. a. Let S_4 be the sum of the areas under the two parabolas in Figure 4. Show the algebra necessary to get the formula for S_4:

$$S_4 = \frac{\Delta x}{3}\left[f(x_0) + 4f(x_1) + 2f(x_2) + 4f(x_3) + f(x_4)\right].$$

   b. Let S_n be the sum of the areas under $n/2$ parabolas. Write a formula for S_n.

5. a. Repeat Problem 2 using Simpson's Rule, again using the table to organize your data.

   b. What value of n did you need for each method to get the answer to the desired accuracy? Which method needed the smallest value of n? (We call this the fastest method.) Which needed the largest? (This is the slowest method.)

   c. Recall that $\Delta x = \frac{b-a}{n}$. It is known that the error in these approximations is roughly proportional to $(\Delta x)^k$, where k is a positive integer. Using your table, find the value of k that you think works for the Riemann sum approximation. Repeat for the Trapezoid Rule and then for Simpson's Rule.

6. The previous problems have been artificial, since we were easily able to compute the integral exactly. As stated in the beginning, we often use numerical integration when we cannot apply the Fundamental Theorem of Calculus. Let us now investigate $\int_0^2 \sqrt{1 + 9x^4}\, dx$, an integral that we cannot do exactly. This integral arises in the calculation of the length of the curve $y = x^3$, where $0 \leq x \leq 2$.

   a. Approximate $\int_0^2 \sqrt{1 + 9x^4}\, dx$ using both the Trapezoid Rule and Simpson's Rule. Experiment with different values of n until you are convinced that your answers are accurate to 0.001.

   b. How did you decide when to stop? How do you get a feel for the accuracy of your answer if you do not have the exact answer to compare it to?

   c. Which method seems the fastest?

Functions given by data

Problem 6 illustrates the use of numerical integration to approximate $\int_{a}^{b} f(x)dx$ when it is difficult or impossible to find an antiderivative for f in terms of elementary functions. In applications it is often the case that functions are given by tables or by graphs, without any formulas attached. For these functions, we only know the function value at specified points. Numerical integration is ideally suited for integrating this type of function. Notice that in this situation we cannot possibly use the Fundamental Theorem of Calculus.

7. A map of an ocean front property is drawn in Figure 5. What is its area?

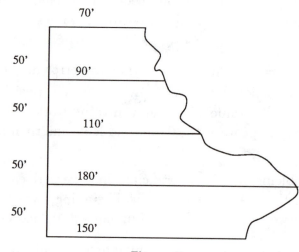

Figure 5

8. The data given below are adapted from the August, 1991 issue of *Road & Track* magazine. They give the velocity $v(t)$ of the $239,000 Lamborghini Diablo at time t seconds.

t	$v(t)$
0	0 mph
1	14 mph
2	27 mph
3	40 mph
4	53 mph
5	64 mph
6	70 mph
7	77 mph
8	84 mph
9	90 mph
10	96 mph

Let $x(t)$ denote the distance the car travels at time t, $0 \leq t \leq 10$. Find $x(10)$. Discuss what method you used and how good you think your answer is.

Further Exploration

9. If $f''(x) > 0$ for all x in $[1, 4]$, would T_{12} be larger or smaller than the actual value of $\int_1^4 f(x)dx$? Why?

10. Although the Trapezoid Rule is usually an improvement over the left and right Riemann sums, it is related to them. Show that the Trapezoid Rule is the average of the left-hand and right-hand Riemann sums. You may do this either algebraically or geometrically.

11. If $f(x)$ is a linear function, it is easy to see that $T_1 = \int_a^b f(x)\,dx$. Similarly, if $f(x)$ is a quadratic, $S_2 = \int_a^b f(x)\,dx$. It is surprising, however, that $S_2 = \int_a^b f(x)\,dx$ if f is a cubic even though the approximating function is not an exact fit. Show that $S_2 = \int_0^1 x^3\,dx$.

Notes to Instructor   Lab 18: Numerical Integration

Principal author: Anita Solow

Scheduling: This lab should occur after the definition of the integral, Riemann sums, and the Fundamental Theorem of Calculus. It could be used either before or after studying techniques of integration.

Computer requirements: Built-in functions for computing Riemann sums, Trapezoid Rule, and Simpson's Rule (given below).

Comments: This lab emphasizes the geometry of the Trapezoid Rule and Simpson's Rule. It tries to explain two situations where numerical integration is desirable, namely, when the antiderivative is difficult to find or when the function is defined by data only. The lab does not discuss the theoretical error bounds of the methods. In the examples in the lab, the students should notice that Simpson's Rule converges more quickly than the Trapezoid Rule.

If the lab is too long, the instructor may choose to omit either the Trapezoid Rule or Simpson's Rule.

Comments on implementing the lab: It is strongly recommended that the instructor program these numerical integration techniques into the computer for students to use. There is nothing to be gained by having students type these in themselves.

In Derive, the following functions will calculate the left-hand Riemann sum, the right-hand Riemann sum, Trapezoid Rule and Simpson's Rule, respectively.

```
f(x) :=
L(a,b,n) := (b-a)/n * sum(f(a + (k-1)*(b-a)/n), k, 1, n)
R(a,b,n) := (b-a)/n * sum(f(a + k*(b-a)/n), k, 1, n)
T(a,b,n) := (b-a)/(2*n) * (f(a) + f(b)
     + 2*sum(f(a + k*(b-a)/n), k, 1, n-1))
S(a,b,n) := (b-a)/(3*n) * (f(a) + f(b)
     + 4*sum(f(a + (2*k-1)*(b-a)/n),k,1,n/2)
     + 2*sum(f(a + 2*k*(b-a)/n), k,1,(n/2)-1))
```

In Maple, the following built-in procedure will work.

```
with(student);
rightsum(f(x), x=a..b, n);
value(");
evalf(");
```

The `rightsum` command produces an unevaluated sum. The last two lines produce a decimal answer. Similarly, in place of `rightsum`, you can use `leftsum`, `trapezoid`, or `simpson`. Alternatively, one can define these directly. You will need to define the function `f` before you can use the commands. Following is the code for the left Riemann sum. The others are similar.

```
left := proc(a,b,n)
   local k,h;
   h := evalf((b-a)/n*sum(f(a+(k-1)*(b-a)/n),k = 1 .. n));
end;
```

In Mathematica, use the following commands:

```
L[f_,a_,b_,n_] := ((b-a)/n)*Sum[f[a + (k-1)*(b-a)/n],{k,1,n}]
R[f_,a_,b_,n_] := ((b-a)/n)*Sum[f[a + k*(b-a)/n],{k,1,n}]
T[f_,a_,b_,n_] := ((b-a)/(2*n)) * (f[a] + f[b]
     + 2*Sum[f[a + k*(b-a)/n], {k, 1, n-1}])
S[f_,a_,b_,n_] := ((b-a)/(3*n)) * (f[a] + f[b]
     + 4*Sum[f[a + (2*k-1)*(b-a)/n],{k,1,n/2}] +
     + 2*Sum[f[a + 2*k*(b-a)/n)], {k,1,(n/2)-1}])
```

To calculate a left-hand Riemann sum of f over the interval $[0, 2]$ using four subintervals, first define the function f and then type `L[f,0,2,4]`. Finish by using `N[%]` to evaluate the result in decimal form.

For programs to perform Riemann sums, the Trapezoid Rule and Simpson's Rule for both the Texas Instrument and Casio graphing calculators, see *Graphing Calculator Manual for Calculus*, by Charlene E. Beckmann and Ted Sundstrom, Addison-Wesley Publishing Company.

Lab 19: Becoming Secure with Sequences

Goals

- To begin thinking about the qualitative behavior of sequences.

- To gain familiarity with some important but non-obvious limits.

- To develop a feel for various rates of growth of sequences that diverge to infinity.

Before the Lab

Let $\{a_n\}$ be a sequence of real numbers. For the purposes of this lab, we distinguish four cases: If $\lim_{n \to \infty} a_n = \infty$, we say $\{a_n\}$ *diverges to infinity*; if $\lim_{n \to \infty} a_n = -\infty$, we say that $\{a_n\}$ *diverges to negative infinity*; if $\lim_{n \to \infty} a_n = c$, a finite real number, we say that $\{a_n\}$ *converges to c*; and in all other cases, we say that $\{a_n\}$ *diverges by oscillation*.

In this lab, you will gain insight into the behavior of $\{a_n\}$ by looking at the way the ordered pairs (n, a_n) lie when plotted in the plane.

1. Before you begin the lab, decide which of the four types of behavior applies to each of the following sequences.

a. $a_n = \dfrac{n}{\sqrt{n}}$

b. $a_n = \dfrac{n + (-1)^n}{n}$

c. $a_n = \dfrac{2^n + n^3}{3^n + n^2}$

d. $a_n = \dfrac{(-1)^n(n-1)}{n}$

106

In the lab

2. *A Friendly Warm-Up.* Use the computer to plot the ordered pairs (n, a_n) for the first 50 terms of each of the four sequences in Problem 1. You may need to adjust the horizontal scale to accommodate all fifty points on your screen, and you may need to adjust the vertical scale so that what you see is consistent with what you expected. Use the same graphical analysis to investigate the behavior of each of the following sequences. Be sure to keep track of the results.

    a. $a_n = \dfrac{\ln n}{\sqrt{n}}$

    b. $a_n = \dfrac{n^{10}}{2^n}$

    c. Experiment with several other values of k for sequences of the form $a_n = \dfrac{n^k}{2^n}$. Does changing the value of k seem to affect the value of the limit? As you change the value of the exponent k you may need to make changes in the vertical scale in order to see all the points of the sequence.

3. $\{r^n\}$ — *the complete story.* In Problem 2c you experimented with modifying the exponent. In this problem you will modify the base.

    a. Write the first few terms of the sequence $\{r^n\}$ for $r = 1$ and $r = -1$. (No need for the machine on this one.) Which does not converge?

    b. Determine the behavior of $\{r^n\}$ for other values of r, both positive and negative, keeping track of the behavior that you observe: divergent to infinity or negative infinity, convergent to a finite real number (which ?), or divergent by oscillation.

    c. Tell the whole story. In your written report for this lab, give a complete description of all possible types of behavior for the sequence $\{r^n\}$ as the value of r ranges over all real numbers. State clearly — with specific examples — which values of r give rise to which type of behavior.

4. *The race to infinity.* Most of the problems in this lab involve the relative rates at which sequences of positive terms diverge to infinity. We clarify the notion of "relative rates" in the following formal definition:

    Let $\{a_n\}$ and $\{b_n\}$ be sequences of positive real numbers that diverge to infinity. We say that $\{a_n\}$ is *strictly faster* than $\{b_n\}$ if and only if $\lim\limits_{n \to \infty} \dfrac{a_n}{b_n} = \infty$, and that $\{a_n\}$ is *strictly slower* than $\{b_n\}$ if and only if $\lim\limits_{n \to \infty} \dfrac{a_n}{b_n} = 0$. If $\lim\limits_{n \to \infty} \dfrac{a_n}{b_n}$ is non-zero and finite, we say that $\{a_n\}$ is *comparable* to $\{b_n\}$.

On the basis of your earlier work in this lab, which is strictly faster, $\{\sqrt{n}\}$ or $\{\ln n\}$? Are $\{2^n\}$ and $\{n^2\}$ comparable? What about $\{2n - 3\}$ and $\{3n - 2\}$? Explain.

5. *Slowing things down.* In this problem, you will slow down the divergence to infinity of a sequence in two different ways: first by taking the logarithm and then by taking the n-th root of each term.

   a. $\{\ln a_n\}$: The sequence $\{2^n\}$ is strictly faster than $\{n\}$, but taking the logarithm of each of its terms results in a sequence that is comparable to n. (Why?) Find another sequence that is strictly faster than $\{n\}$ but which, when slowed down in this way, becomes strictly slower than $\{n\}$. Assemble graphical evidence to support your example.

   b. $\{a_n^{1/n}\}$: Taking the n-th root of the n-th term in the sequence $\{2^n\}$ results in the constant sequence $\{2\}$. (That's really putting on the brakes!) Discuss what happens to the sequence $\{n\}$ when you slow it down in this way. Does $\{n^{1/n}\}$ have a finite limit? What is it? Give graphical evidence to support your claim.

6. $\left\{\left(1 + \dfrac{r}{n}\right)^n\right\}$— *variations on a theme.*

   a. It is a well-known and important fact that $\left\{\left(1 + \dfrac{r}{n}\right)^n\right\}$ converges to e^r. By letting $r = 1$, $\ln 2$, and $\ln 3$, assemble evidence in support of this fact.

   b. Investigate the behavior of two related sequences $\left\{\left(1 + \dfrac{1}{n^2}\right)^n\right\}$ and $\left\{\left(1 + \dfrac{1}{\sqrt{n}}\right)^n\right\}$. Discuss how the sequences differ in their behavior as n gets large. What do you think accounts for this difference?

Further Explorations

7. In Problem 6 you dealt with sequences of the type $\left\{\left(1 + \dfrac{1}{a_n}\right)^{b_n}\right\}$ where both a_n and b_n diverged to infinity. In the first part of Problem 6, a_n and b_n diverged to infinity at the same rate; in the second part, they diverged to infinity at different rates. Try to find a precise connection between the limit of the sequence and the relative rates at which a_n and b_n diverge to infinity.

8. a. Apply the n-th root procedure to slow down the sequence $\{n!\}$. Graph the result. It still diverges to infinity, but at a slower rate. The points of the sequence should appear to lie along a straight line of constant slope.

    b. Use this to find a simpler sequence that is comparable to $\{n!^{1/n}\}$.

    c. Get an estimate for the numerical value of the slope. Look up Stirling's formula for $n!$ in your textbook or in a book of tables. Explain the connection between this formula and your estimate for the numerical value of the slope.

Notes to Instructor   Lab 19: Becoming Secure with Sequences

Principal author: John Fink

Scheduling: Just after limits of sequences have been introduced and students have had a chance to work with sequences that exhibit the types of behavior which this lab explores: non-convergence, convergence to a finite real number, and divergence to infinity.

Computer or calculator requirements: Ability to plot points of a sequence $\{a_n\}$ as discrete ordered pairs (n, a_n) in the plane. In the absence of this ability, treating n as a continuous variable is possible, but this gives a misleading impression of what is being assessed.

Comments: The purpose of this lab is twofold: to get students to think qualitatively — instead of computationally — about sequences; and to cultivate some sensitivity to the rate at which a sequence diverges to infinity. In Problems 4 and 5, this notion of "rate of growth" is introduced, and in Problem 6, various rates are contrasted.

Although we suggest that you try all the labs before giving them to your students, it is especially important here. Some computer systems may be too slow to do the computations in an acceptably small amount of time. In Problem 2, we ask for 50 terms of a sequence. If your system is slow, you could reduce the number of terms, or you could have the student evaluate the sequence at a few selected values of n instead of plotting all the terms.

Comments on implementing the lab: The lab will be most effective if some graphical representation of the terms of a sequence is available. For example, the first fifty terms in the sequence whose nth term has been defined to be `a[n]` or `a(n)` can be plotted as follows:

In Mathematica:

```
ListPlot[Table[a[n],{n,1,50}]]
```

In Derive: If your machine is slow, it might help to set the precision to Approximate mode for this lab.

```
Author Vector([n,a(n)],n,1,50) approXimate Plot
```

In Maple: Type `plot(map(op,[[n,a(n)]$n=1..50]),style=POINT);`

Lab 20: Getting Serious about Series

Goals

- To understand series convergence in terms of limits of partial sums.

- To explore the family of p-series.

- To investigate the growth of the partial sums for the harmonic series.

- To study the alternating harmonic series and what can happen when terms are rearranged.

In the Lab

In trying to attach meaning to an infinite series $\sum_{k=1}^{\infty} b_k$ of real numbers b_k, it is natural to study the behavior of the *partial sums* $S_N = \sum_{k=1}^{N} b_k$. If $\lim_{N \to \infty} S_N$ exists, we say that the series *converges* and we define the value of the limit to be the *sum* of the infinite series. Otherwise, we say the series *diverges*.

1. *A friendly warm-up.* Consider the geometric series $\sum_{k=0}^{\infty} r^k$, where the real number r gives the ratio between successive terms. Note that the series starts with $k = 0$ rather than $k = 1$.

   a. Let $r = \frac{4}{5}$ and use the computer to investigate the partial sums of the series $\sum_{k=0}^{\infty} (\frac{4}{5})^k$. As you evaluate partial sums with more and more terms, what appears to be the limiting value? Confirm that this value for the sum of the infinite series agrees with the well-known formula for the sum of a geometric series whose ratio r has absolute value less than 1.

   b. Let $r = 1.01$ and use the computer to study the partial sums. What do you conclude about the convergence or divergence of the geometric series in this case? Explain.

   c. Now let $r = 1$ and $r = -1$. Discuss the convergence or divergence of the series in these cases. Do you really need to use a computer in part *c* or even in part *b*? Explain.

d. When r is negative in a geometric series we get an example of an *alternating series*. Explain why this is an aptly chosen description. Give examples of two alternating geometric series, one convergent and one divergent.

e. Investigate the behavior of the geometric series for various values of the ratio r. Write a clear and complete statement summarizing your results.

2. *The p-series.* Now consider the *p-series* $\sum_{k=1}^{\infty} \dfrac{1}{k^p}$, where p can be any positive real number.

a. For $p = .5$ and for $p = 2$, write the first five terms of the series in your notebook. Then by studying partial sums on the computer, argue that one of these series converges and one diverges. Do not get too ambitious here on the number of terms, N, in your partial sums; depending on your computer system, you may be in for an annoying wait. Give an estimate of the sum for the convergent series. [Obscure hint indicating a beautiful and mysterious fact: Multiply the partial sums by 6 and take the square root of the results.]

b. Select two more p values, one of which gives a convergent p-series and the other a divergent one. Give explanations and evidence for your conclusions.

c. When $p = 1$ we get a special and important case of the p-series known as the *harmonic series*, $\sum_{k=1}^{\infty} \dfrac{1}{k}$. Write out and then compute the sum of the first four terms of the harmonic series by hand. Then use the computer to estimate the partial sums for $N = 4$, $N = 128$, and $N = 1024$.

d. Based upon your results of parts a and b, formulate a preliminary conjecture about convergence and divergence of the p-series in terms of the positive real number p. Ignore the $p = 1$ case for now. That is the subject of the next problem.

3. *The harmonic series.* It is a remarkable and important fact that the harmonic series diverges. Were you persuaded otherwise by the fact that its terms tend to 0? Any straightforward attempt to compute this sum by a computer, no matter how large or powerful, will lead to the incorrect conclusion that the series converges. (See Problem 3e below.) In this problem we will use the computer to obtain a valuable hint about how to prove that the harmonic series diverges.

a. Building on what you did in Problem 2c, make a table of values for the partial sums of the harmonic series with $N = 4, 8, 16, 32, 64$, and 128. Confirm from your table that each time enough successive terms are added for N (the number of terms in the partial sum) to reach the next power of two, the partial sum increases by an amount exceeding $\frac{1}{2}$.

b. Assuming that this subtle but systematic process of increasing by $\frac{1}{2}$ persists for arbitrarily large N, argue, with no help from the computer, that the values of the partial sums will eventually exceed 14. Give a value of N for which the partial sum exceeds 14. Justify your answer.

c. Try to prove what you observed in part a: that by successively adding enough terms to any partial sum, we can further increase its value by at least $\frac{1}{2}$. Hint: Consider

$$\tfrac{1}{5} + \tfrac{1}{6} + \tfrac{1}{7} + \tfrac{1}{8}$$

and

$$\tfrac{1}{9} + \tfrac{1}{10} + \tfrac{1}{11} + \tfrac{1}{12} + \tfrac{1}{13} + \tfrac{1}{14} + \tfrac{1}{15} + \tfrac{1}{16}.$$

How many terms are in the first sum and what is the smallest term? What about the second sum? Finally, think similarly about the general case,

$$\frac{1}{2^k + 1} + \frac{1}{2^k + 2} + \cdots + \frac{1}{2^{k+1}}.$$

Now put this all together in the clearest and most persuasive argument you can make, and you will have a proof that the harmonic series diverges.

d. Now, refine your answer to Problem 2*d* with a more definitive statement about convergence and divergence of *p*-series.

e. (Optional, but highly recommended for computer buffs) Write a paragraph or two attempting to justify, in your own words, the claim made above that any attempt to evaluate the harmonic series by direct numerical summation of terms on any computer using any amount of time will result in a finite sum and lead to the faulty conclusion that the series converges. Hint: The sum it "converges" to will depend on the precision being used in the calculation.

4. *The alternating harmonic series.*

a. If we change the signs of all the even numbered terms in the harmonic series, we get the *alternating harmonic series,* $\displaystyle\sum_{k=1}^{\infty} \frac{(-1)^{k+1}}{k}$. Write out the first four terms in the alternating harmonic series and compute the fourth partial sum by hand. Make sure you see how the $(-1)^{k+1}$ factor causes the signs to alternate.

b. We state without proof the fact that the alternating harmonic series converges. Use the computer to estimate the sum of this series. Hint: Compare your result with $\ln 2$.

Further Exploration

5. *Rearrangement of terms in the alternating harmonic series.*

a. In many ways and in many situations, convergent infinite sums behave like finite ones. The infinite has its share of surprises, however, and one such surprise is that by suitably changing the order of the terms in the alternating harmonic series, the rearranged series can be made to converge to any real number we please. Consider the following rearrangement:

$$1 - \tfrac{1}{2} - \tfrac{1}{4} + \tfrac{1}{3} - \tfrac{1}{6} - \tfrac{1}{8} + \tfrac{1}{5} - \tfrac{1}{10} - \tfrac{1}{12} + \cdots .$$

Write the next six terms in the series to make sure you see the pattern. Use the computer to make your best guess about the sum of this series. Explain what you did. Can you confirm your result algebraically? Hint: Write the series as

$$\left(1 - \tfrac{1}{2}\right) - \tfrac{1}{4} + \left(\tfrac{1}{3} - \tfrac{1}{6}\right) - \tfrac{1}{8} + \cdots$$

and compare with the original alternating harmonic series.

b. Find a rearrangement of terms of the alternating harmonic series that converges to a negative value. Explain your rearrangement rule and provide some evidence that the sum is indeed negative. The computer will help here, but think before you compute.

6. *Another alternating series.* Consider the alternating 2-series $\displaystyle\sum_{k=1}^{\infty} \frac{(-1)^{k+1}}{k^2}$.

Estimate the sum of this series. If terms in this alternating series are rearranged, does it seem to affect the sum? Justify your answer. Compare your results here with your results in Problem 5. Why do you think this alternating series is more "well-behaved" than the alternating harmonic series?

7. *A baseball joke.* What is $\displaystyle\sum_{k=1}^{\infty} \frac{1}{k^{\text{world}}}$?

Notes to Instructor   Lab 20: Getting Serious about Series

Principal author: Ed Packel

Scheduling: This lab can be scheduled to introduce or reinforce the ideas related to the convergence of an infinite series of constants.

Computer or calculator requirements: Computing finite sums.

Comments: This lab is intended to let students understand and explore series of constants in terms of their associated sequences of partial sums. After a warm-up with geometric series, students are challenged to explore the *p*-series and to guess which values of *p* give convergent series and which do not. Help is given for the case of the celebrated harmonic series in the form of hints for exhibiting the pattern utilized in the standard divergence proof. Also, the alternating harmonic series is introduced. The section of further exploration considers the problem of rearranging terms in alternating series.

Comments on implementing the lab: In Problem 1, some computer algebra systems will sum the full geometric series directly, with no need for partial sums. It may be more instructive for students to approach the series via partial sums.

In Problem 2, some computer algebra systems (like Mathematica) are capable of directly approximating the infinite series, while others (Derive, for instance) require successive evaluation of some wisely chosen partial sums. Either approach is fine as long as the equivalence of series convergence and partial sum convergence is firmly set in the student's mind. Some systems (Maple is one) may confuse the issue on the $p = 2$ partial sums by giving too sophisticated an answer.

The harmonic series is notorious for how slowly it diverges. Care should be taken not to waste excessive computer and human time. See R. P. Boas, Jr. and J. W. Wrench, Jr., "Partial sums of the harmonic series," *American Mathematical Monthly*, 78 (1971), 864–869. The value $N = 1024$ in Problem 2 should be small enough not to tax most computer systems.

Lab 21: Limit Comparison Test

Goal

- To examine the Limit Comparison Test and its proper use.

In the Lab

A fundamental idea in the study of series is that of comparison. We look at $\sum_{n=1}^{\infty} \frac{n-2}{n^2}$ and feel that it should behave pretty much like $\sum_{n=1}^{\infty} \frac{n}{n^2} = \sum_{n=1}^{\infty} \frac{1}{n}$, so we try to formalize this idea. We write $\frac{n-2}{n^2} < \frac{n}{n^2} = \frac{1}{n}$, and then we realize that this inequality goes the wrong way; $a_n < b_n$ and $\sum_{n=1}^{\infty} b_n$ diverges. From this information, one is not allowed to conclude anything about $\sum_{n=1}^{\infty} a_n$.

Still, we have the feeling that we should be able to use what we know about $\sum_{n=1}^{\infty} \frac{1}{n}$. And our feeling is right! It's just that we need a variation on the comparison test. The Limit Comparison Test gives us what we need.

Limit Comparison Test: If $\{a_n\}$ and $\{b_n\}$ are sequences of positive numbers such that $\lim_{n \to \infty} \frac{a_n}{b_n}$ is a positive real number (greater than zero, but finite), then $\sum_{n=1}^{\infty} a_n$ converges if and only if $\sum_{n=1}^{\infty} b_n$ converges.

The goal of this lab is to give you a feel for how you choose a comparison series in order to use the Limit Comparison Test.

1. Use the Limit Comparison Test to determine if $\sum_{n=1}^{\infty} \frac{n-2}{n^2}$ converges or diverges.

2. Consider the series $\displaystyle\sum_{n=1}^{\infty} \frac{1}{n^{3/2} + n}$.

   a. Use your computer to graph the functions $f(x) = \dfrac{1}{x^{3/2} + x}$ and $g(x) = \dfrac{1}{x^{3/2}}$ on the same axes. Set the scale on the horizontal axis so the x-values go from 1 to 100. We are interested in comparing the functions for large values of x, and when x is large, both $f(x)$ and $g(x)$ are quite small. Thus you should set a scale on the vertical axis to view the y-values between 0 and 1. Sketch the graphs in your notebook.

   b. Use your computer to plot the function f defined in part a, together with $h(x) = \frac{1}{x}$ on the same axes. Sketch the results in your notebook. You may have to experiment to find a range of y-values that will enable you to see both functions on the same axes.

   c. Use your computer to calculate $\displaystyle\lim_{x \to \infty} \frac{f(x)}{g(x)}$ and $\displaystyle\lim_{x \to \infty} \frac{f(x)}{h(x)}$ for the functions in parts a and b above. Note the results.

   d. Look at the results of parts a through c and the statement of the Limit Comparison Test. Should you compare the series $\displaystyle\sum_{n=1}^{\infty} \frac{1}{n^{3/2} + n}$ with $\displaystyle\sum_{n=1}^{\infty} \frac{1}{n^{3/2}}$ or with $\displaystyle\sum_{n=1}^{\infty} \frac{1}{n}$? Explain your choice after considering which term dominates the denominator for large values of n.

   e. Does the series $\displaystyle\sum_{n=1}^{\infty} \frac{1}{n^{3/2} + n}$ converge? Use the Limit Comparison Test and your knowledge of p-series (series of the form $\sum \frac{1}{n^p}$) to justify your conclusion.

3. Let us consider a different series, $\displaystyle\sum_{n=1}^{\infty} \frac{1 + n^2}{n^3 + n^2}$. Experiment with graphs and limits until you find a p-series with the appropriate value of p so you can apply the Limit Comparison Test to this new series. Use the same techniques that you did in answering Problem 2 above. Note your attempts and why you rejected the failures, as well as why you chose the p you did. Does this series converge or diverge? Why?

4. Look next at $\displaystyle\sum_{n=1}^{\infty} \frac{n + \sqrt[3]{n}}{n^{7/3} + n^2}$. Again experiment with graphs and limits in order to choose a suitable comparison series. This time consider the dominant term in the numerator as well as the dominant term in the denominator. If you ignore the other terms, what p-series does this series most resemble? Does the series converge or diverge?

5. All of the examples have been series of the form $\displaystyle\sum \frac{n^a + n^b}{n^c + n^d}$. To what series of the form $\displaystyle\sum \frac{1}{n^p}$ would you compare $\displaystyle\sum \frac{n^a + n^b}{n^c + n^d}$ in order to use the Limit Comparison Test? How do you decide on your answer?

Further Exploration

6. a. Adapt the ideas you developed in the lab to determine whether the series $\displaystyle\sum_{n=0}^{\infty} \frac{1}{3^n - 2^n}$ converges. Find an appropriate geometric series and use the Limit Comparison Test

   b. Also determine whether $\displaystyle\sum_{n=0}^{\infty} \frac{2^n + 5^n}{3^n + 6^n}$ converges. Try to generalize your result to series of the form $\displaystyle\sum_{n=0}^{\infty} \frac{a^n + b^n}{c^n + d^n}$ for positive constants a, b, c, and d.

   c. What series would you use to test the convergence of $\displaystyle\sum_{n=0}^{\infty} \frac{1}{2^n + n}$? What is the result of the Limit Comparison Test for this series?

   d. How would you handle the series $\displaystyle\sum_{n=0}^{\infty} \frac{3^n + n^5}{n^3 + 5^n}$? Does this series converge?

7. Why does the Limit Comparison Test insist that $\displaystyle\lim_{n\to\infty} \frac{a_n}{b_n}$ be not only finite but positive? What can you conclude if $\displaystyle\lim_{n\to\infty} \frac{a_n}{b_n} = 0$?

Notes to Instructor   Lab 21: Limit Comparison Test

Principal authors: Bonnie Gold and Robert Messer

Scheduling: This lab is designed to be used just after the Limit Comparison Test has been introduced in class.

Computer or calculator requirements: Most of this lab can be done with a graphics package alone. A computer algebra system might be useful for calculating some limits at infinity, but these can easily be calculated by hand.

Comments: Students often have difficulty applying the Limit Comparison Test because they have trouble choosing a suitable comparison series. This lab helps them experiment both formally (via limits) and graphically to see when two series can be compared.

Lab 22: Approximating Functions by Polynomials

Goals

- To introduce the idea of one function being a good approximation to another.

- To prepare students for work on Taylor polynomials and Taylor series.

In the Lab

Polynomials can be easily evaluated at any point and their integrals are easy to find. This is not true of many other functions. Thus, it is useful to find polynominals that are good approximations to other functions.

In this lab we will find polynomials that approximate the exponential function. This function is important in mathematics and frequently appears in models of natural phenomena (population growth and radioactive decay, for instance). In these situations we need an easy way to approximate e^x for all values of x, not just for integers and simple fractions. Also, integrals involving the exponential function are important in statistics. For example, $\dfrac{1}{\sqrt{2\pi}} \displaystyle\int_0^{.5} e^{-x^2}\, dx$, which calculates the probability of a certain event that follows the "bell-shaped curve" of the normal distribution, simply cannot be evaluated in terms of the usual functions of calculus.

We will rely on the computer's ability to evaluate and graph the exponential function in order to determine polynomials that appear to be good approximations to this function. We will also use our polynomial approximations to compute integrals involving the exponential function.

1. We begin with a constant function that best approximates e^x near $x = 0$. Why is the graph of $y = 1$ the best constant approximation to the graph of $y = e^x$ near $x = 0$? That is, why would $y = 2$ or $y = -1$ be a worse approximation to $y = e^x$ near $x = 0$? Let us denote this polynomial approximation of degree zero by p_0.

2. Now we want to add a first degree term to p_0 to find a polynomial of the form $1 + ax$ that best approximates e^x near 0. Use your computer to graph $y = e^x$ and several candidates such as $y = 1 + .5x$, $y = 1 + .9x$, and $y = 1 + 1.2x$ on the same axes. Keep in mind that you are looking for the value of a so that $1 + ax$ best approximates e^x near 0. Thus you should favor a line that follows along the curve $y = e^x$ right at 0. You may need to change your scale to decide which line is better.

   In your notebook record which lines you tried and explain the criteria you used in choosing the line that gives the best approximation. Let $p_1(x) = 1 + ax$ denote your choice of the line that best approximates e^x near 0.

3. Next, find the second degree term bx^2 to add to p_1 to get a quadratic polynomial $p_2(x) = 1 + ax + bx^2$ that best approximates e^x near 0. Try to get a parabola that follows along the graph of $y = e^x$ as closely as possible on both sides of 0. Again, record the polynomials you tried and why you finally chose the one you did.

4. Finally, find a third degree term cx^3 to add to p_2 to get a cubic polynomial $p_3(x) = 1 + ax + bx^2 + cx^3$ that best approximates e^x near 0. This may not be so easy; you may have to change scale several times before you see why one polynomial is better than another.

5. Now that you have a polynomial that approximates e^x, try evaluating $p_3(.5)$ as a computationally simple way of estimating $e^{.5}$. How close is the polynomial approximation to the value of $e^{.5}$ as determined by a calculator or computer? Which is larger? How does the error at other points of the interval $[0, .5]$ compare with the error at $x = .5$? If you cannot distinguish between the graphs of $y = p_3(x)$ and $y = e^x$, you may want to plot the difference $y = e^x - p_3(x)$ with a greatly magnified scale on the y-axis.

6. Let us return to the problem of computing a definite integral such as $\int_0^{.5} e^{-x^2}\, dx$ for which the integrand does not have an antiderivative in terms of elementary functions. Since $p_3(x)$ approximates e^x, we can use $p_3(-x^2)$ to approximate e^{-x^2}.

   a. Evaluate $\int_0^{.5} p_3(-x^2)\, dx$ as an approximation to $\int_0^{.5} e^{-x^2}\, dx$.

   b. Use the numerical integration command on your computer to approximate $\int_0^{.5} e^{-x^2}\, dx$. How does this compare with your answer from part a?

7. An analytical method for approximating a function near a point leads to what are known as *Taylor polynomials*. The Taylor polynomial of degree n is determined by matching the values of the polynomial and its first n derivatives with those of the function at a particular point.

   a. Make a table to compare the values of p_3 and its first three derivatives with the values of e^x and its derivatives, all evaluated at $x = 0$. How close was your polynomial p_3 to being a Taylor polynomial?

   b. Determine the cubic Taylor polynomial for the exponential function. To do this, adjust the four coefficients so the values of the Taylor polynomial and its first three derivatives match those of e^x at $x = 0$. Plot this polynomial and your polynomial p_3. Compare how close they are to the graph of $y = e^x$ near 0.

Further Exploration

Suppose we want a polynomial that approximates a function over some **fixed** interval, rather than in some vaguely defined interval "near" a given point. This is important, for example, when we approximate a definite integral of a function by the integral of the polynomial over the interval. There are many different methods for approximating a function over a given interval. Therefore do not worry if your answers to Problem 8 are quite different from those obtained by other students. The point of this problem is for you to think about criteria you might use to judge the quality of an approximation. You may want to take a course in numerical analysis for further information about the surprising variety of techniques for approximating functions.

8. a. Determine the constant polynomial, the first degree polynomial, and the quadratic polynomial that you feel best approximate e^x on the whole interval $[-1, 1]$. Record your attempts. Do all three polynomials have the same constant term? Did you change the coefficient of x when you went from the straight line to the parabola? What criteria are you using to decide if one polynomial is a better approximation than another?

   b. Compare your criteria for deciding when a polynomial is a good approximation near a point, and your criteria for deciding when a polynomial is a good approximation over a given interval. How and why are your criteria different?

Notes to Instructor   Lab 22: Approximating Functions by Polynomials

Principal author: Bonnie Gold

Scheduling: This lab is intended to precede any work with Taylor polynomials. Students need to be familiar with the exponential function.

Computer or calculator requirements: This lab requires a graphics package that allows you to compare several plots on the same screen. Problem 6 requires a numerical integrator to compare $\int_0^{.5} p_3(-x^2)\,dx$ with $\int_0^{.5} e^{-x^2}\,dx$.

Comments: Students are not expected to get the Taylor polynomial exactly by comparing graphs, but they should come reasonably close. The cubic Taylor polynomial for e^x is $1 + x + \frac{x^2}{2} + \frac{x^3}{6}$. Students are unlikely to get $\frac{1}{6}$ as the coefficient for the x^3 term, but $1 + x + \frac{x^2}{2} + .1x^3$ or $1 + x + \frac{x^2}{2} + .2x^3$ are quite good. For $p_2(x)$, even $1 + x + .4x^2$ isn't bad; but they should not get $1 + x + x^2$.

The aim of this lab is to have students think about what is meant by a good approximation. They should realize that this is a nontrivial concept and appreciate the simplicity of the formula for the Taylor polynomials when this is discussed in class.

Exploratory Problem 8 can lead to a spirited class discussion, if you ask students to give their results and justify them.

Lab 23: Radius of Convergence for Power Series

Goals

- To obtain graphical evidence for the interval of convergence of a power series.

- To estimate the radius of convergence visually for selected power series.

- To see the radius of convergence as a function of the center about which the Taylor series is expanded.

Before the Lab

A series of the form $\sum_{n=0}^{\infty} a_n(x - c)^n$ is called a *power series centered at c*. Associated with any such series is its *interval of convergence*: for all values of x interior to this interval, the series converges; for all values of x outside this interval, the series diverges. For the purposes of this lab, we ignore the behavior of the series for values of x at the endpoints of this interval. The constant c lies at the center of this interval. The distance from c to either endpoint is called the *radius of convergence* of the series. Your textbook illustrates how to use the root test and the ratio test to determine this radius precisely for certain series. The purpose of this lab is to give you a graphical context for these analytical facts.

1. Use the techniques developed in your textbook to compute the radius of convergence for each of the series below:

    a. $\sum_{n=0}^{\infty} 2^n(x-2)^n$

    b. $\sum_{n=0}^{\infty} \dfrac{(-1)^n(x-1)^{n+1}}{n+1}$

    c. $\sum_{n=0}^{\infty} \dfrac{(-1)^n x^{2n+1}}{2n+1}$

One of the most important features of power series is their ability to represent familiar functions over their intervals of convergence. If the function f has derivatives of all orders at some point c, then we can form the series $\sum_{n=0}^{\infty} a_n(x-c)^n$ where $a_n = \dfrac{f^{(n)}(c)}{n!}$. This is called the *Taylor series for f centered at c*. For most functions f that you are familiar with (and for all but one of the functions in this lab), the Taylor series will converge to $f(x)$ for all values of x within the interval of convergence. The partial sums for this series are polynomials of degree n:

$$p_n(x) = a_0 + a_1(x-c) + a_2(x-c)^2 + \cdots + a_n(x-c)^n.$$

In most of the standard cases, for values of x within the interval of convergence, $p_n(x)$ will be a good approximation for $f(x)$ when n is large enough. For values of x outside the interval of convergence, $p_n(x)$ is usually a poor approximation for $f(x)$, no matter how large n is.

For example, the Taylor series for the familiar natural logarithm function centered at 1 is $\sum_{n=0}^{\infty} \dfrac{(-1)^n(x-1)^{n+1}}{n+1}$. You computed the radius of convergence of this series in Problem 1b above. Figure 1 superimposes the graph of $y = p_{16}(x)$ onto the graph of $y = \ln x$.

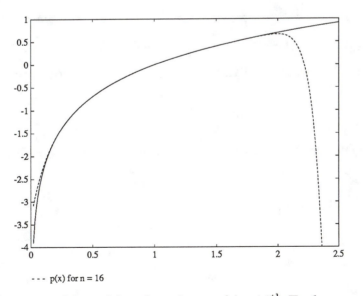

--- p(x) for n = 16

Figure 1: The natural logarithm function and its 16th Taylor approximation

Notice that $p_{16}(x)$ represents $\ln x$ pretty well over the interval of convergence, but that it pulls away sharply outside of that interval.

2. Highlight the interval on the x-axis in Figure 1 on which p_{16} seems to agree with the logarithm function.

The Taylor series for the arctangent function centered at 0 is $\displaystyle\sum_{n=0}^{\infty} \frac{(-1)^n x^{2n+1}}{2n+1}$.
You computed its radius of convergence in Problem 1c. The figure below superimposes the graph of $y = p_{15}(x)$ onto the graph of $y = \arctan x$.

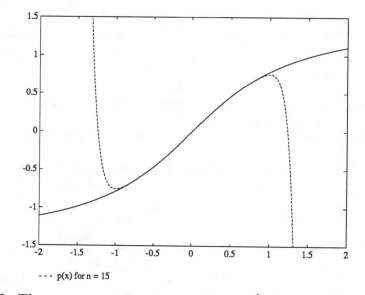

--- p(x) for n = 15

Figure 2: The arctangent function and its 15th Taylor approximation

3. Highlight the interval on the x-axis in Figure 2 on which p_{15} seems to agree with the arctangent function. How does this compare with the radius of convergence you computed in Problem 1c?

In the Lab

In the problems that follow, you will be asked to produce Taylor polynomials and to superimpose their graphs onto those of the functions they represent. We will consistently use p_n to denote the nth degree Taylor polynomial for a function f centered at a point c. In each case, you will be asked to determine the radius of convergence for the associated Taylor series and to support your conclusion with graphical evidence similar to that above.

4. Let $f(x) = \ln x$ and $c = 2$.

    a. The Taylor series for f centered at $c = 2$ is $\ln(2) + \displaystyle\sum_{n=1}^{\infty} \frac{(-1)^{n-1}(x-2)^n}{n2^n}$.
        Write the first four terms of this series to make sure you see the pattern.

    b. Use your computer to produce p_3 centered at 2 for $f(x) = \ln x$. Check that the coefficients agree with the ones you recorded for p_3 in part a. In the viewing rectangle $[0,5] \times [-10,2]$ plot both $y = \ln x$ and $y = p_3(x)$. Also have your computer generate p_6 and add its graph to the screen. Copy your graphs onto your data sheet. Be sure to label each curve and to indicate the units in the x and y directions. On the basis of this graphical evidence, highlight the interval of convergence in your sketch.

    c. Compute the radius of convergence for this series and discuss the extent to which it agrees with the graphical evidence.

5. Again let $f(x) = \ln x$. This time we will vary the value of c.

    a. Use your computer to find p_7 centered at $c = 5$. Plot it along with $y = \ln x$ in the rectangle $[0, 15] \times [-10, 5]$. Note the apparent interval of convergence. Plot p_n for several larger values of n to get a feeling for the behavior of these polynomials over the interval of convergence. When you have obtained a picture that illustrates your idea of the interval of convergence, copy it onto your data sheet, making sure to label each curve. Explain why this picture supports your conclusion.

    b. By now you have seen that the radius of convergence changes when the point c changes. Experiment with several other positive values of c. How does the radius of convergence seem to depend upon the value of c?

6. Let $f(x) = \sqrt[3]{1+x}$ and consider the Taylor series for f centered at $c = 0$, at $c = 1$, and at $c = -2$.

   a. Use the techniques of Problem 4 to estimate the radius of convergence for the Taylor series of f when centered at $c = 0$, at $c = 1$, and at $c = -2$.

   b. How does the radius of convergence seem to depend upon c for this function? What similarities between the graph of $y = \sqrt[3]{1+x}$ near $x = -1$ and the graph of $y = \ln x$ near $x = 0$ might account for the dependence you observe?

7. Let $f(x) = \arctan x$.

   a. The arctangent function and all of its derivatives are perfectly well-behaved for every real value of x. Nevertheless, the radius of convergence is finite. When $c = 0$, you have already seen that the radius is 1. Use the techniques established above to estimate the radii of convergence when $c = 1$ and when $c = 2$.

   b. Estimate the radii of convergence for other values of c between -2 and 2. Record your data as points on a graph showing the radius of convergence as a function of the location of the center.

Further Exploration

8. This problem introduces a function whose Taylor series converges everywhere, but not to the value of the function. Let

$$f(x) = \begin{cases} e^{-1/x^2}, & \text{for } x \neq 0, \\ 0, & \text{for } x = 0. \end{cases}$$

   a. Use the computer to sketch a graph of this function on the interval $[-1, 1]$. For which values of x does $f(x)$ appear to be 0? For which values of x does $f(x)$ actually equal 0?

   b. It appears that $f'(0) = 0$. Sketch the graph of f'. What does $f''(0)$ appear to be? Use your computer to derive and then sketch some of the higher derivatives of f. What does $f^{(n)}(0)$ appear to be in each case?

   c. In this case, appearances are not deceiving. It is a remarkable fact that f has derivatives of all orders at 0, and $f^{(n)}(0) = 0$ for all n. Given this fact, what is the Taylor series for f centered at $c = 0$? What is the interval of convergence for this series? For which values of x does the series converge to $f(x)$?

9. Obtain a formula in terms of n for the coefficient of $(x - c)^n$ in the Taylor series for $f(x) = \ln x$. Use this formula to determine the radius of convergence for this Taylor series. How does the radius of convergence depend on c? How does this analytical result compare with the graphical results you found in Problem 4?

10. Although the derivatives of $f(x) = \arctan x$ are well-behaved for all real numbers, the same cannot be said for all complex numbers. For which complex numbers x is the function $f(x) = \dfrac{1}{1 + x^2}$ undefined? How might this account for your results in Problem 6?

Notes to Instructor   Lab 23: Radius of Convergence for Power Series

Principal author: John Fink

Scheduling: This lab can occur at any time after students have learned how to use the root or ratio tests to compute the radius of convergence for power series.

Computer requirements: Ability to return Taylor polynomials for functions specified by the user. Ability to graph two functions in the same viewing rectangle.

Comments: The purpose of this laboratory is to provide compelling graphical support for the radius of convergence for a power series. Behavior at the endpoints can be discussed later, after the students have gained some visual understanding of this concept.

Comments on implementing the lab: The student will need to be able to produce Taylor polynomials of degree n centered at c for various functions f, and then plot them along with f in the same viewing rectangle. The examples below show how to do this in three different systems for the fifth Taylor polynomial of $\arctan x$ at $c = 1$, using a $[-1,3] \times [-2,2]$ plotting window.

In Mathematica:

```
f[x_] := ArcTan[x]
taylor[f_, c_, n_] := Normal[Series[f[x],{x,c,n}]]
t = taylor[f,1,5]
Plot[{f[x],t}, {x,-1,3}, PlotRange->{-2,2}]
```

In Maple:

```
f(x) := arctan(x);
plot({convert(taylor(f(x),x=1,5),polynom),f(x)},
     x=-1..3,y=-2..2);
```

In Derive, to keep f from being prematurely bound, assign it to the empty expression before defining p(c,n).

```
f(x) :=
p(c,n) := Taylor(f(x),x,c,n)
f(x) := atan(x)
p(1,5)
```

Then press `Expand`. Switch to the `Plot` screen. Derive is very slow at plotting graphs that lie outside the viewing rectangle. Adjusting the viewing rectangle can have dramatic impact on the plotting speed. `Move` the cross to $(1,0)$, `Center`, and then `Scale` x at .5 units per tick mark. Finally, `Plot` both expressions.

Some packages compute complex roots for expressions involving $x^{1/3}$, so the graph may be missing for $x < 0$. This happens in Derive, for example, unless you overrule the default branch of the cube root function by giving the commands `Manage`, `Branch`, `Real`. Check your package out on Problem 5, and omit the series centered at $c = -2$ if necessary.

Warn students not to get too carried away with high degree Taylor polynomials. Students will rapidly discover that the computer takes a fair amount of time to expand or plot these. Furthermore, the numerical errors that accumulate when these are evaluated may lead to graphs with quite a bit of static.

Lab 24: Polar Equations

Goals

- To match elements of the polar graph $r = r(\theta)$ with corresponding values of θ.

- To develop mathematical formulas for slope and angle in polar coordinates, and to use them to address questions that lie beyond the reach of strictly graphical analysis.

- To match aspects of the graph $y = f(x)$ in rectangular coordinates against aspects of the graph $r = f(\theta)$ in polar coordinates.

Before the Lab

The purpose of this lab is to give you some experience in working with the graphs of equations defined by polar functions. We will use the fact that any polar function can be written in parametric form. If the polar coordinates of a point P are (r, θ), then the rectangular coordinates of P will be $(x, y) = (r \cos \theta, r \sin \theta)$. This leads to the standard parametrization of a polar curve $r = r(\theta)$ by

$$x(\theta) = r(\theta) \cos \theta,$$

and

$$y(\theta) = r(\theta) \sin \theta.$$

1. Consider the graph of a function defined parametrically by $x = x(t)$ and $y = y(t)$. The slope of curve at point $(x(t), y(t))$ is given by $\dfrac{y'(t)}{x'(t)}$. Use this result and the standard parametrization of a polar curve $r = r(\theta)$ given above to show that the slope of a polar graph at the point $(r(\theta), \theta)$ is given by

$$\frac{r'(\theta) \sin \theta + r(\theta) \cos \theta}{r'(\theta) \cos \theta - r(\theta) \sin \theta}.$$

In the lab, a polar curve will be given as the graph of the function $r(\theta)$ in polar coordinates. You will need to be able to match points on the curve with their corresponding values of θ in the interval $[a, b]$. For example, points in the first quadrant will correspond to values of θ in the interval $[0, \frac{\pi}{2}]$ when $r(\theta)$ is positive on this interval.

2. What condition on r will force points in the first quadrant to correspond to values of θ between π and $\frac{3\pi}{2}$? Explain.

3. Given a polar graph of a function r, how can you determine which values of θ correspond to points $(r(\theta), \theta)$ where the curve crosses the x-axis?

In the Lab

4. *Flowers.*

   a. Use your graphing utility to verify that the polar graph of $r(\theta) = \sin 3\theta$, for θ in the interval $[0, 2\pi]$ is shaped like a flower with three petals. Would a domain smaller than $[0, 2\pi]$ produce the same graph? Explain. What values of θ correspond to the petal in the first quadrant?

   b. Sketch the polar graph of $r(\theta) = \sin 2\theta$, θ in $[0, 2\pi]$. How many petals do you see? Which values of θ correspond to the petal in the first quadrant? Could you have gotten the same graph from a smaller domain for θ this time?

   c. Experiment with the polar graphs of $r(\theta) = \sin k\theta$ for other integer values of k. Make a conjecture about the connection between the integer k and the number of petals in the polar graph.

5. *Limaçons.* Polar graphs of the functions $r(\theta) = 1 + k \sin \theta$ are called *limaçons*, French for snails, possibly because of their vague resemblance to the shape of these animals for certain values of k. See what you think of the resemblance by sketching the polar graph of $r(\theta) = 1 + 2 \sin \theta$, for θ in $[0, 2\pi]$ (These were probably named by the same kind of people who call the big dipper Ursa Major).

   a. Your sketch for $k = 2$ should show a smaller loop inside a larger loop. Which values of θ correspond to this smaller loop?

   b. Sketch the limaçons for $k = .75$ and $k = .25$ to see other possible shapes. One should be dented, the other egg-like. Sketch the limaçons for several additional values of k. Besides being looped, dented, or egg-like, did you find any other shapes occuring as limaçons?

6. *Spirals.* If $r(\theta)$ is positive and is strictly increasing as a function of θ, the polar graph of r will be shaped like a spiral. For the spirals considered here, we will be mainly interested in their behavior as they cross the positive x-axis, for positive values of θ.

   a. *Logarithmic Spirals.* Sketch the polar graph of the function $r(\theta) = \ln \theta$ for values of θ in the interval $[1, 8\pi]$. For which values of θ does this curve cross the positive x-axis? What are the values of x at these points?

b. *Arithmetic Spirals.* Find a function r whose polar graph is a spiral which meets the positive x-axis at precisely the integer values: 1, 2, 3, ... Begin your search by listing the values of θ for which the spiral will cross the positive x-axis, and then consider what the value of $r(\theta)$ will need to be at these points. When you think you've got the right function, sketch your graph for values of θ in $[0, 8\pi]$ to be sure.

c. *Exponential Spirals.* Find a function r whose polar graph meets the positive x-axis at precisely the powers of 2 : 1, 2, 4, 8, ... Proceed as you did with arithmetic spirals, listing the values of θ for which the spiral crosses the positive x-axis, and then considering what the value for $r(\theta)$ will need to be at these points. Sketch your graph for values of θ in $[0, 8\pi]$ to confirm that you have the right function.

In the next five parts of this problem, we consider the angle at which a spiral meets the x-axis.

d. Use the formula you derived in Problem 1 to show that the slope of the line tangent to a spiral given by the function r at the points $(r(\theta), \theta)$ where it crosses the x-axis is given by $\dfrac{r(\theta)}{r'(\theta)}$.

e. Your sketch of the logarithmic spiral probably looked perpendicular to the x-axis at most of the points where it crossed the x-axis, but looks can be deceiving. Use the formula that you just derived to help explain the truth.

f. The sketch of your arithmetic spiral probably appears to meet the positive x-axis in the same angle at each point of intersection. Does it really? Explain.

g. The sketch of your exponential spiral probably also always appears to intersect the positive x-axis at the same angle. This time looks do not deceive. Explain.

h. Find a function r which gives a spiral that always crosses the x-axis in a 45° angle.

Further Exploration

7. *Limaçons Revisited.*

   a. Select two values of k, one corresponding to a looped limaçon and one to a non-looped one. Use these values to sketch the graphs of $y = 1 + k \sin x$ in rectangular coordinates. Note those points (if there are any) where the curves cross the x-axis. Is there anything about these rectangular graphs that would account for the shape of the associated limaçons?

   b. In Problem 5*b*, you found the shapes that the graph of a limaçon could have. Explain how you know that you found all of the shapes and determine exactly which values of k correspond to each shape. Analyzing graphs will not be sufficient to answer this question. You will need to use your formula for the slope of the tangent line to find those places on the limaçon where the tangent is horizontal.

Notes to Instructor   Lab 24: Polar Equations

Principal author: John Fink

Scheduling: This is an introductory laboratory on graphing in polar coordinates. It assumes that students know how to make the correspondence between points in the plane and their polar coordinates. It can be used any time after a discussion of parametrizing a curve in the plane by rectangular coordinate functions $x(t)$ and $y(t)$.

Computer or calculator requirements: Ability to produce plots of curves given parametrically or in polar form.

Comments: One of the families of curves students are asked to study are the limaçons, given as the graphs of $r(\theta) = 1 + k\sin\theta$. They come in three shapes: looped, dented, or oval. The graph of $y = 1 + k\sin x$ in rectangular coordinates can be used to determine which values of k correspond to the looped shape, but a more subtle analysis is required to distinguish between the values of k for dented and oval limaçons. This problem was derived from a longer problem written by Arthur Sparks and Gary Klett at a National Science Foundation Summer Workshop at St. Olaf College.

The answers to Problem 6h $(r(\theta) = ce^{\theta})$ may not fit on the screen. The students will therefore have to rely on the mathematics they have developed to convince themselves that the curves have the desired property.

With the exception of these large spirals there should be no surprises in using the computer to produce the polar graphs of the functions r found in this lab.

Comments on implementing the lab:

In Derive: In the 2D-Plot Window, use `Options-Type-Polar`, and enter a and b as prompted.

In Mathematica: Use the command
`ParametricPlot[r[t]{Cos[t],Sin[t]},{t,a,b},AspectRatio->Automatic]`

In Maple: Use the command `plot([r(t),t,a..b], coords=polar);`

All graphics calculators graph either polar functions or parametric functions, or both, by changing the mode.

Lab 25: Differential Equations and Euler's Method

Goals

- To illustrate differential equations as a modeling tool and as a major application of calculus.

- To solve differential equations numerically by Euler's method.

- To investigate how the accuracy of approximate solutions depends on the step size.

Before the Lab

One of the major applications of calculus involves the formulation and solution of differential equations that arise in many fields of study. An example from physics, well supported by experimental data, is a model for radioactive decay. This model says that if $A(t)$ denotes the level of radioactivity of a substance at time t, then the rate of change in $A(t)$ is proportional to $A(t)$ itself. Let r denote the proportionality constant. The value of this decay constant will depend on the particular radioactive material. Because we are modeling decay, dA/dt will be negative. By requiring $r > 0$, we get the differential equation

$$\frac{dA}{dt} = -rA(t).$$

If we know the value of A at $t = 0$, often written $A(0) = A_0$, we have what is known as an *initial-value problem*. General theorems from the theory of differential equations assure us that virtually any reasonable initial-value problem will have a solution and that the solution will be unique. This will certainly be true for the differential equations we encounter in this lab.

1. a. Think about a family of functions whose derivatives are just themselves times a multiplicative constant. Now guess a solution A for the above differential equation that models radioactivity decay.

   b. Check that any constant multiple of a solution to the above differential equation is also a solution. Use the initial condition $A(0) = A_0$ to determine the constant multiple. Put your results together to record the unique function A that satisfies the initial-value problem.

c. Try out your solution to the differential equation with decay constant $r = 2$ and initial value $A(0) = 10$. Compute $A(.1)$, $A(.2)$ and $A(1)$ as predictions of the levels of radioactivity at times $t = .1$, $.2$, and 1.

As is the case with indefinite integrals, many differential equations have solutions that cannot be expressed in terms of familiar functions. Fortunately, solutions to differential equations can be approximated to almost any desired degree of accuracy by a variety of interesting and widely used techniques. Such an approximation is called a *numerical solution* to the differential equation. In this lab we will investigate a very simple and natural approximation technique, known as Euler's method, for producing a numerical solution. Even though more sophisticated methods are used in practice, Euler's method serves as an excellent introduction to the numerical solution of differential equations.

Let us consider the initial-value problem given by

$$\frac{dy}{dt} = f\big(t, y(t)\big), \qquad y(t_0) = y_0.$$

Here y is an unknown function of t that we seek as the solution to the differential equation, and f is a known function that depends on both t and y. The function y is a solution in the sense that the derivative of y evaluated at t is precisely $f\big(t, y(t)\big)$, the value we obtain from the function on the right hand side of the differential equations when we plug in t and $y(t)$. Of course the solution y must have the value y_0 at t_0, thus also satisfying the initial condition.

Euler's method gives an approximate solution for t values near t_0 by using $\dfrac{y(t + h) - y(t)}{h}$ as an estimate for $y'(t)$. The increment h is called the *step size*. Values of h close to zero generally give better estimates for $y'(t)$ and hence better approximations.

2. a. Estimate $y'(t_0)$ by the difference quotient based on the increment in t from t_0 to $t_1 = t_0 + h$. Convince yourself that this approximation converts the differential equation into the approximate equality $\dfrac{y(t_1) - y(t_0)}{h} \approx f(t_0, y_0)$.

b. Use the approximation in part a to show that, at t_1, the exact value of the solution $y(t_1)$ can be approximated by $y_1 = y_0 + h f(t_0, y_0)$.

c. Now let $t_2 = t_1 + h$ and $t_3 = t_2 + h$ so that in general $t_k = t_{k-1} + h = t_0 + kh$. Repeat the argument above to obtain an approximation to $y(t_2)$. In addition to approximating $y'(t_1)$ by the difference quotient, you will need to approximate $f\big(t_1, y(t_1)\big)$ by $f(t_1, y_1)$. This is reasonable provided f is continuous and y_1 is close to $y(t_1)$.

d. Carry out the approximation argument one or more steps until you are convinced that the above process can be continued to approximate $y(t_{n+1})$ by

$$y_{n+1} = y_n + hf(t_n, y_n), \qquad n = 0, 1, 2, \ldots .$$

This is the defining equation of Euler's method.

e. What is the function f in the differential equation for radioactive decay? Do not be alarmed that t does not appear in the formula for f. Keep in mind that r is a constant. If there is any chance you will interpret the decay constant as a variable, set it equal to 2 and work with that.

f. Use parts b and c above with step size $h = .1$, decay constant $r = 2$, and initial value $A_0 = 10$ to estimate the solution to the radioactive decay equation at the points $t_1 = .1$ and $t_2 = .2$. Call these values A_1 and A_2. How do they compare with the values $A(t_1)$ and $A(t_2)$ of the true solution that you obtained in Problem 1.

In the Lab

3. Your instructor will provide you with a program for applying Euler's method to obtain an approximate solution of a differential equation.

   a. Test this on the differential equation for radioactive decay with step size $h = .1$, decay constant $r = 2$, and initial value $A_0 = 10$. Do the approximations at $t = .1$ and $t = .2$ agree with your answers in Problem 2f?

   b. Use the program, again with $h = .1$, to estimate the solution at $t = 1$. Compare this estimate with the value of the true solution $A(1)$ obtained in Problem 1c. See if you can get your computer to compare the results of Euler's method at each step along the interval $[0, 1]$ with the true solution. Try to get both a numerical and a graphical comparison.

4. Let us consider the influence of the step size when Euler's method is applied to the differential equation $\dfrac{dA}{dt} = -rA(t)$. As before, take $r = 2$ and $A_0 = 10$. Try a variety of step sizes, some less than .1 and some greater than .1 to study the relation between the accuracy and the step size. Collect and organize enough data to make a convincing argument for your conclusions. What price do you pay for increased accuracy?

5. Let us now try Euler's method in a slightly more complicated situation. Suppose the rate of growth of an algae population depends on the number of algae cells and the temperature. Specifically, suppose a biologist determines that the population $P(t)$ at time t satisfies the initial-value problem

$$\frac{dP}{dt} = .003P(t)\left(2 + 5\sin(\tfrac{\pi}{12}t)\right), \qquad P(0) = 45000.$$

The factor $2 + 5\sin(\tfrac{\pi}{12}t)$ has a period of 24 to model the daily temperature fluctuation when t is measured in hours.

a. Use Euler's method to obtain a numerical approximation to the solution over a time interval of four days. Experiment to determine a reasonable value of the step size.

b. Present your results numerically and graphically.

c. Describe the qualitative features of your results that might be of interest to a biologist.

6. Listed below are three differential equations with initial conditions. One has an easily guessed solution, one has the solution $y(t) = t\cos t$, and one has no solution in terms of known functions.

   i. $\dfrac{dy}{dt} = \cos t - y\tan t, \qquad y(0) = 0$

   ii. $\dfrac{dy}{dt} = t\cos(yt), \qquad y(0) = 0$

   iii. $\dfrac{dy}{dt} = t^2, \qquad y(0) = 0$

a. Solve the equation which has the easily guessed solution by making a good guess and showing that it satisfies the differential equation.

b. For the one with solution $y(t) = t\cos t$, show that $y(t)$ is indeed a solution. Also solve it numerically on an appropriate interval and compare your results.

c. For the remaining differential equation, solve it numerically on an interval of your own choosing.

Further Exploration

7. What do you think happens if the step size in Euler's method is negative? Make a conjecture based on the geometric ideas behind the formula for Euler's method. Test your conjecture on the radioactive decay equation and other initial value problems from this lab.

8. As mentioned earlier, Euler's method is just a naïve beginning of the numerical fun one can have in solving differential equations. Two of the many numerical methods for solving initial-value problems are called the Improved Euler method and the Runge-Kutta method. Look up one of these methods in a textbook on differential equations or numerical analysis.

    a. Give an intuitive explanation of how the method works.

    b. Write or acquire a program to implement it.

    c. Test it out on some of the equations in this lab.

    d. Compare it to Euler's method for speed and accuracy.

Notes to Instructor   Lab 25: Differential Equations and Euler's Method

Principal author: Ed Packel

Scheduling: This lab may be used to reinforce the topic of differential equations or as an introduction to the numerical solution to differential equations.

Computer requirements: Iteration is the main tool required for this lab. It is used in a program to implement Euler's method. The ability to plot a discrete set of points is highly desirable. This enables the students to present the results of Euler's method graphically. Here are some specific ways to implement a program for Euler's method.

Derive: Make sure students select `Options`, `Input`, `Word` to allow identifiers of more than one letter. `Author` the generic function `f` of two variables and then `Author` the rule for `n` steps of Euler's method with initial condition (`t0,y0`) and step size `h`.

```
f(t,y):=
euler(t0,y0,h,n):=
       iterates(v+[h,h*f(element(v,1),element(v,2))],v,[t0,y0],n)
```

The output is a vector of ordered pairs, $[t_k, y_k]$ for $k = 0, 1, 2, \ldots, n$. The `Plot` command then plots these points. Be sure to adjust the scales on the axes appropriately.

Mathematica: The following code will work:

```
f[t_,y_]:=Cos[t]-y*Tan[t]
g[{u_,v_}]:={u+h,v+h*f[u,v]}
h=desired stepsize
euler[t0_,y0_,n_]:=NestList[g,{t0,y0},n]
```

The output will be a list of ordered pairs of the form

$$\{\{t_0, y_0\}, \{t_1, y_1\}, \{t_2, y_2\}, \ldots, \{t_n, y_n\}\}.$$

Graphs can be obtained by applying ListPlot to the output of euler.

Maple: First define the function `f(t,y)`. Then the following definition of the function `euler` will produce a list of the form $t_0, y_0, t_1, y_1, \ldots, t_n, y_n$.

```
euler := proc(t0,y0,h,n)
    t.0 := t0;
    y.0 := y0;
    Table := t.0,y.0;
```

```
    for i from 0 to n do
        t.(i+1) := t.i+h;
        y.(i+1) := y.i+h*f(t.i,y.i);
        Table := Table,t.(i+1),y.(i+1)
    od:
  end;
```

To graph the output type plot([Table], style=LINE);

Lab 26: Shapes of Surfaces

Goals

- To understand the shape of a surface by analyzing its x and y cross sections.

- To find points which are critical points of both the x and y cross sections.

- To analyze graphically the nature of such points.

In the Lab

In this lab you are to use your computer to produce graphs of functions of two variables. For each problem, you are to plot a graph of the given function over the specified rectangular domain. Feel free to change the viewing position of the graph in order to see the important features of the surface. You will need a printed copy of each surface on which to mark your answers.

1. Let $f(x, y) = 4 - x^2 - y^2$, over the domain $-8 \le x \le 8$ and $-8 \le y \le 8$.

    a. Mark and label the point $(8, -8, -124)$ on the graph. Explain how we know that this point is on the graph of the function.

    b. Using only algebra, not calculus, explain why $(0, 0, 4)$ is the highest point on the graph.

    We will now examine x and y cross sections of this surface. An $x = k$ cross section is the intersection of a plane $x = k$ with the surface $z = 4 - x^2 - y^2$. To find its equation we set $x = k$: $z = 4 - k^2 - y^2 = -y^2 + (4 - k^2)$. This is a parabola in the yz coordinate system in the plane $x = k$. We can define the y cross sections in an analogous manner. The graph that the computer draws is made up of x and y cross sections.

    c. Mark and label the $x = 0$ cross section on the graph. Next, write down its equation, graph it by hand on a yz coordinate system, and describe it geometrically. Use calculus to find its unique critical point. Is this point a maximum of the cross section, a minimum, or neither? Also mark this point on the graph of the surface.

    d. Repeat part c for the $y = 0$ cross section, this time graphing its equation on a xz coordinate system.

    e. Repeat part c for both the $x = -4$ and $y = -4$ cross sections.

144

Notice that the $x = 0$ and $y = 0$ cross sections intersect at the point $(0, 0, 4)$, which you showed in part *b* to be the highest point on the surface. The $x = -4$ and $y = -4$ cross sections intersect at $(-4, -4, -28)$, which is not a particularly interesting point on the surface. Also notice where the critical points that you found in parts *c*, *d*, and *e* occurred, and start thinking about how you could use these critical points to distinguish between the interesting point $(0, 0, 4)$ and the uninteresting point $(-4, -4, -28)$.

2. Let $f(x, y) = x^2 - y^2$, where $-4 \leq x \leq 4$ and $-4 \leq y \leq 4$.

   a. Mark and label the points $(0, 0, 0)$ and $(2, -2, 0)$ on the graph of the function.

   b. Using only algebra, explain why $(0, 0, 0)$ is neither the highest nor lowest point on the graph. (Although $(0, 0, 0)$ is not a high point or a low point, it is of special interest; it is called a *saddle point* because of the saddle-like shape of the surface near it.)

   c. Repeat Problem 1c for the $x = 0$ cross section of this graph.

   d. Repeat Problem 1c for the $y = 0$ cross section of this graph.

   e. Repeat Problem 1c for the $x = 2$ and $y = -2$ cross sections of this graph.

From Problems 1 and 2, you have probably begun to guess at a general pattern concerning critical points of cross sections and their relation to the interesting points on the graph that are called maximum points (high points), minimum points (low points), and saddle points. Try to state this general pattern and then apply it to Problem 3. Your description of the general pattern should help you find the maximum, minimum, and saddle points. (Caution: Your description is likely to be not entirely correct at this stage, since you haven't seen many graphs yet. But you will have a chance to improve your description after Problems 3 and 6.)

3. In this problem we will examine the surface defined by the function $f(x, y) = 3x - x^3 + 12y - y^3$, where $-4 \leq x \leq 4$ and $-4 \leq y \leq 4$.

   a. This surface has one maximum, one minimum, and two saddle points. Mark them as best you can on the surface and estimate their (x, y) coordinates.

   b. To understand why we get one maximum, one minimum and two saddle points, we need to understand the x and y cross sections of this surface. Use calculus to find the critical points of each x cross section and of each y cross section. Use these to pinpoint the exact locations of the four interesting points. Furthermore, identify what kind each is by examining the shapes of the two cross sections through each of these four points.

4. Was your guess at a general pattern successful in helping you answer Problem 3, or do you need to improve your guess? State as clearly as you can what you now believe the general pattern to be.

Further exploration

5. In order to do this problem, you will need to use the level curves (or contour plots) for each of the surfaces defined in Problems 1, 2 and 3.

   a. On each picture, mark all of the points that are critical points of both the $x = k$ and $y = k$ cross sections of the surface.

   b. Describe the shape of the level curves near the maximum and minimum points.

   c. Describe the shape of the level curves near the saddle points.

   d. Figure 1 is a graph of several level curves of a function f. Using your results to parts *b* and *c*, mark the approximate position of each of its maximum, minimum, and saddle points. Determine how many extremum (*i.e.* maximum or minimum) and how many saddle points this surface has within the given domain.

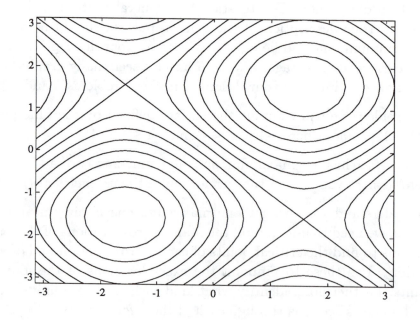

Figure 1: Level curves for f

6. Let $f(x,y) = x^4 - xy + y^4$.

    a. Show that the origin is the unique critical point of both the $x = 0$ and $y = 0$ cross sections. Show it is a minimum of both cross sections.

    b. Locate the point $(0,0)$ on the level curves of f, given in Figure 2. The level curves seem to indicate that $(0,0)$ is not a minimum point for f. What aspects of the graph indicate this?

    c. In order to confirm that the origin is not a minimum of the surface, analyze f along the $y = x$ cross section. Along this cross section, $z = 2x^4 - x^2$. The point $(0,0)$ is a critical point along this cross section. However, it is a maximum, not a minimum. Use calculus to verify this fact. The moral of the story is that just because a point is a minimum along both the x and y cross sections, it need not be a minimum of the surface.

    d. Do you need to correct your guess of the general pattern you gave in Problem 4? If so, do it now.

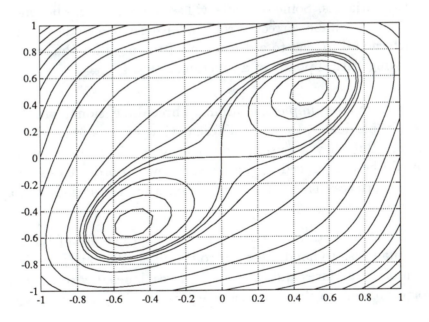

Figure 2: Level curves for $f(x,y) = x^4 - xy + y^4$

Notes to Instructor   Lab 26: Shapes of Surfaces

Principal authors: Anita Solow and Eugene Herman

Scheduling: This lab can be scheduled at the beginning of the study of functions of several variables. It does not require the use of partial derivatives.

Computer requirements: 3-dimensional graphics are required. It is assumed that the students will be able to get paper copies of the graphs. (See below for additional comments.) Level curves are needed for Problem 5 in the Further Exploration section. These are supplied below.

Comments: This lab is designed to prepare students for the study of partial derivatives. We expect many students to perceive the following. At any maximum, minimum, or saddle point P, the x and y cross sections through P both have critical points at P. Thus, students anticipate one of the main theorems concerning functions of two variables. Some students will see more patterns beyond this central one, not all of which will be correct. Problem 6 is designed to disabuse them of one such false pattern and to lead to a discussion of the need for the second derivative test for classifying critical points of a function.

Comments on computer implementation: Three-dimensional graphics packages vary greatly. In particular, they use different viewpoints and mesh sizes for their default values. But they all use x and y cross sections to graph the surfaces. Students should be encouraged to change the views in order to see the important features on the surfaces. These are often hidden when the surface is graphed using the default parameters.

If it is difficult for your students to produce paper copies of the graphs of the surfaces, you may wish to hand these out to the class. We have supplied you with a graph of each of the surfaces and their level curves, labeled with the number of the problem corresponding to that function. Feel free to copy these for your students' use.

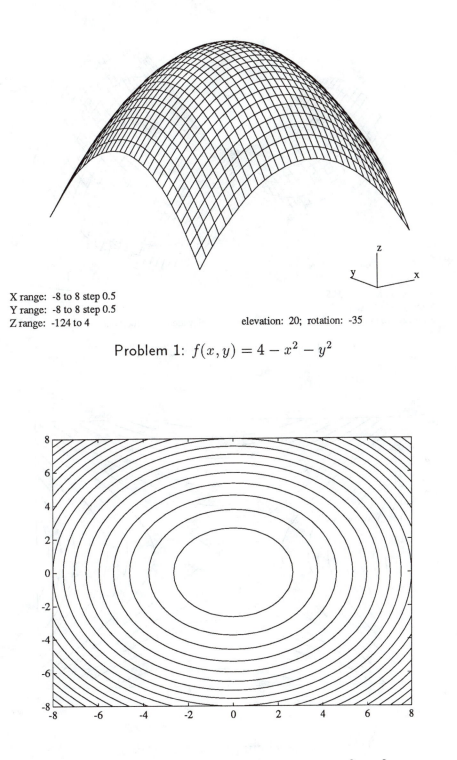

X range: -8 to 8 step 0.5
Y range: -8 to 8 step 0.5
Z range: -124 to 4   elevation: 20; rotation: -35

Problem 1: $f(x, y) = 4 - x^2 - y^2$

Problem 1: Level curves for $f(x, y) = 4 - x^2 - y^2$

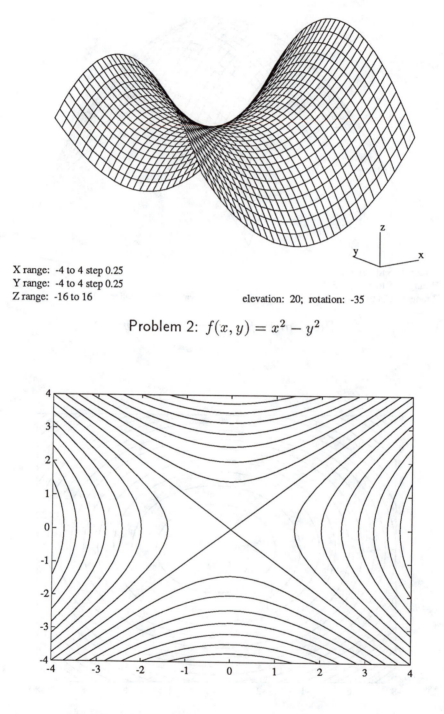

X range: -4 to 4 step 0.25
Y range: -4 to 4 step 0.25
Z range: -16 to 16

elevation: 20; rotation: -35

Problem 2: $f(x,y) = x^2 - y^2$

Problem 2: Level curves for $f(x,y) = x^2 - y^2$

X range: -4 to 4 step 0.25
Y range: -4 to 4 step 0.25
Z range: -68 to 68

elevation: 20; rotation: -35

Problem 3: $f(x, y) = 3x - x^3 + 12y - y^3$

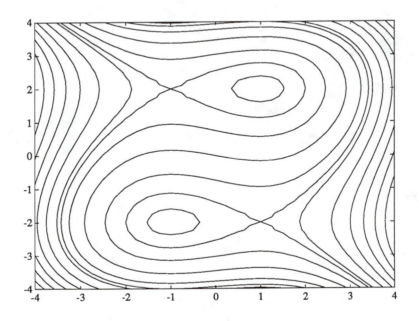

Problem 3: Level curves for $f(x, y) = 3x - x^3 + 12y - y^3$

Syllabi for Calculus I & II

The following syllabi were developed by first soliciting suggestions from the faculty members of the twenty-six participating schools, returning all of the suggestions to all of the contributors and asking them to find as much common ground as possible, and by forming a committee of active contributors to combine the suggestions into a cohesive document.

The committee has worked with two goals in mind: that the course should be lean and that it have unifying themes. The first goal is as simple to state as it is difficult to achieve: identify the basic ideas that should go into the core of any one-year calculus course. We decided from the outset to restrict the syllabi to ideas that would fit into 32 to 35 class meetings in each of two terms. We thought that such a syllabus would leave time for:

- dwelling on topics that instructors particularly want to emphasize

- drawing from the resource collections to include large- scale applications, individual projects, and cultural background

- training students in the use of electronic learning aids

- testing

For the second goal, while the first calculus course has the unifying ideas of differentiation and integration linked by the Fundamental Theorem of Calculus, many instructors find the second course a conglomeration of ideas and techniques that students find unrelated. Our suggestion is to build the second course around the ideas of precision and approximation. That is, to investigate methods that produce exact solutions, see when these methods fail or do not apply, and when that is the case to find ways to get approximate solutions and to estimate the size of the errors made. This structure would give the second course a measure of cohesiveness while simultaneously emphasizing the importance of making approximations, something that present courses do not often do but, we think, ought to. As an alternative to this course, Calculus IIA, we provide Calculus IIB in which functions of two and three variables are introduced to generalize concepts of the first semester.

As we worked towards the goals, we made some assumptions about the manner in which we thought the courses should be taught. They include:

- The obvious should not be proved. For example, plausibility arguments should suffice for such things as the formula for the derivative of a sum, the First Derivative Test, and the Intermediate Value Theorem.

- Proofs need not be completely rigorous, provided that deficiencies are explained. For example, a full and rigorous proof of the Chain Rule is not necessary in the first course.

- Topics should be brought up only when necessary. For example, continuity does not have to be mentioned until the Mean Value Theorem is discussed.

- Functions should not be defined solely by equations. Tables and, especially, graphs should be used continually in examples and exercises.

- Extra time should be spent on investigating topics more fully, and not on introducing additional topics.

Jean Calloway (Kalamazoo College)
Bonnie Gold (Wabash College)
Harold Hanes (Earlham College)
Paul Humke (St. Olaf College)
Andrew Sterrett (Denison University), Chair

Calculus I: The Derivative and the Integral

Introduction (1 class)

Describe geometrically the two problems that dominate first-year calculus. Begin with a special case of the area problem (for example, find the area bounded by $y = x^2$, $x = a$, and the x-axis) by calculating the limit (intuitively, of course) of upper sums that approximate the area. Then describe the tangent problem (for example, find the slope of the tangent line to $y = x^3/3$ at $x = a$) and find the limit of the appropriate difference quotient. Complete the introduction by promising that there is a relation between these apparently unrelated problems, and mention the roles of Newton and Leibniz. If a computer algebra system is available, the two problems can easily be finished in a single class period.

Functions and Graphs (4)

- definition; domain and range; linear and quadratic functions
- trigonometric functions (sine, cosine, and tangent)
- exponential and logarithmic functions
- composite functions
- functions described by tables or graphs

The emphasis in this part of the course should be on graphing a broader than usual collection of functions, including functions described by tables, so that a rich collection of examples and applications will be available throughout the year. One sequence of assignments could require students to graph a large number of functions by any method, with examples of each type covered so far included each day.

Note that limits and continuity are not included in this section. The idea of a limit is introduced in the next section, when its lack would prevent defining and finding instantaneous rates of change. Similarly, the idea of continuity should not be brought up until it arises naturally.

The Derivative (10)

- average rates of change

- instantaneous rates of change, developed intuitively

- a study of limits, either intuitive or epsilon-delta

- definition and properties of the derivative

- derivatives of polynomials

- derivatives of sines and cosines

- derivatives of exponentials and logarithms

- derivatives of sums, differences, products, and quotients

- the chain rule and inverse functions

In the first class a number of applications where average rates of change arise should be discussed. For example, average velocity, average revenue, slope of a secant line, and other more specialized applications such as the average rate at which a person's body assimilates and uses calcium, the average rate of production at a plant, and so on.

Applications that could be considered include motion on a line, freely falling bodies, and related rates.

Extreme Values (8)

- extreme values; approximate graphical or numerical solutions using a calculator or computer, if available

- Existence Theorem: a function continuous on a closed interval attains maximum and minimum values

- Critical Point Theorem: a function attains extreme values only at critical points

- Concavity Theorem: a function with a positive second derivative is concave up

- First Derivative Test for local extremes

- Second Derivative Test for local extremes

- The Mean Value Theorem

In the first class, a number of interesting word problems whose solutions involve finding an extreme value should be introduced and they should be used to motivate the ideas that follow.

The Mean Value Theorem provides an excellent opportunity to discuss the uses of existence theorems by noting how this one leads to several important results, such as the Monotonicity Theorem and the fact that two functions with the same derivative differ by a constant. Problems like "Find the value of c where..." should not be emphasized.

Applications include the exact solution of earlier word problems and finding more detailed information about graphs of functions.

Antiderivatives and Differential Equations (3)

- antiderivatives and some of their basic properties

- an introduction to differential equations; separation of variables; constants of integration and initial conditions

Possible applications are exponential growth, escape velocity, falling bodies, and Torricelli's Law.

The Definite Integral (6)

- Riemann sums

- limits of Riemann sums

- Integrability Theorem; properties of definite integrals

- The Fundamental Theorem of Calculus

- the derivative of integrals with variable upper limits (in two ways)

Consistent with the introduction to the derivative, the introduction to the definite integral should begin with examples where the function to be integrated is piecewise constant. Students know that distance traveled is the product of velocity and time when velocity is constant but they do not know how to find distance when velocity varies with time. The Riemann sum provides an approximate solution. Other examples and applications include areas, volumes, work, and moments.

The Fundamental Theorem of Calculus provides the opportunity to remind students of their first class, when they were made aware of two important geometric problems and were promised that they would see a relation between them.

Calculus IIA: Exact and Approximate Representation of Numbers and Functions

The theme of Calculus II is the representation of numbers and functions by several methods, both exact and approximate. Topics considered are sequences as functions, improper integrals as limits of sequences of numbers, functions approximated by polynomials or represented as power series, functions described by the behavior of their derivatives, exact and approximate solution of differential equations, and area and volumes considered as integrals. Calculators and computer algebra systems make easy the use of many techniques to find both exact and approximate solutions. The constant interplay of precise and approximate representations should be the focus of this course.

Introduction (1 class)

Introduce several examples that illustrate the role that approximate as well as exact solutions have in mathematics. For example,

- Remind students how to find $\sin 30°$ and indicate how to approximate $\sin 31°$ with a Taylor polynomial. (The use of a computer to graph the sine and its polynomial approximation is particularly effective.)

- Exhibit the integral for arc length (to be derived later) and discuss its limitations and how they might be overcome.

- Indicate graphically how Newton's method generates a sequence of approximations.

- Discuss how to estimate π and the importance of error analysis.

The Definite Integral Revisited (9)

- the definite integral: exact values from the Fundamental Theorem of Calculus

- antiderivatives: using substitution, including trigonometric substitution, and integration by parts

- the definite integral: approximate values by using Riemann sums and by the Trapezoidal Rule, with some error analysis.

New applications of the definite integral, or old applications that require new methods for finding antiderivatives or for evaluating integrals of functions that lack simple antiderivatives, should be used to motivate the ideas and techniques in this section. The approach used to introduce new applications should be consistent with that used earlier, that is, using the integral to add up generic pieces of things rather than presenting formulas to be memorized.

Integration by parts and by substitution are included here because they are important in solving differential equations and in other mathematics courses. Other antiderivatives can be found by using tables or a computer algebra system.

Emphasis should be placed on error analysis, including the graph of an appropriate derivative, of the numerical methods used to approximate definite integrals.

Possible applications include arc length (including curves defined by parametric equations), Buffon's needle problem, surface area, numerical integration of tabular data, and Monte Carlo methods to estimate area.

Sequences and Series of Numbers (10)

Sequences topics:

- infinite sequences as functions

- limits of sequences

- recursively defined sequences

- improper integrals, including l'Hopital's rule

- limits at infinity and the asymptotic behavior of functions

The need to study sequences of numbers should be introduced by an appropriate example. For instance, finding the area bounded by $y = \frac{1}{x}$ and the x-axis over the interval $[1, \infty)$ and the amount of paint it would take to fill the infinite funnel obtained by revolving $y = \frac{1}{x}$ about the x-axis. One or more of the applications listed below could also be used.

Special emphasis should be placed on the rates of growth or convergence of sequences.

Possible applications include compound interest (compounded n times a year or continuously), approximating π or e, Newton's method for finding zeros, repeating decimals as approximations to rationals, and decimal approximations to irrational numbers.

Series topics:

- infinite series

- geometric series

- the nth term test for divergence

- equivalence of series, and the Limit Comparison Test

- the harmonic series and p-series

Motivation for this material can come from the desire to find exact or approximate values of series that result from applications. A key idea is comparing the rate of growth of a series of positive terms with other series whose behavior is known. A starting point is the formula for the sum of a finite geometric series which allows discussion of convergence as a limit of a sequence of partial sums and which leads to conditions under which infinite geometric series converge and diverge. Convergence of a general series can be defined and examples can show what it means for a series to be comparable to a geometric series. The Ratio and Integral Tests can be introduced as "formalized comparisons." Students should be able to compare the rates of convergence and divergence for p-series, geometric series, and series involving factorials. Errors should be estimated in every case where that is possible.

Applications include the distance traveled by a bouncing ball, numerical values of π and e, Zeno's paradox of Achilles and the tortoise, and the multiplier effect in economics.

Sequences and Series of Functions (8)

- The Mean Value Theorem revisited and its second-degree analogue

- Taylor polynomials with remainder theorem

- graphical comparison of a function with its Taylor polynomials; the graph of the error function for a Taylor approximation

- error estimation on intervals

- Taylor series: the general expansion and examples (sine, cosine, exponential, logarithm, the binomial theorem, and so on)

- power series, with the Ratio Test applied to give domains of convergence

- algebraic manipulation and term-by-term integration and differentiation

The emphasis should be on three ideas: that a sequence of functions is an extension of the notion of a sequence of numbers, that sequences of Taylor polynomials are naturally associated with differentiable functions and can be used to approximate them to prescribed tolerances, and that techniques of manipulation acquired earlier still serve in this new environment.

Students could be asked to print a small three decimal table of the sine function as an application.

Series Solutions of Differential Equations (4)

- defining functions with differential equations, for example $y'' + ky = 0$ and $y' = ky$

- solving homogeneous linear second-order equations with constant coefficients using power series

Differential equations were introduced in Calculus I, where equations with separable variables were solved. This section reintroduces the idea and provides a good opportunity to introduce applications that lead to differential equations that give rise to trigonometric and exponential equations.

IIB: Calculus in a Three-dimensional World

The purpose of Calculus IIB is to extend the concepts of derivative and integral to three-space. After revisiting the definite integral in two-space, the concepts of double and iterated integrals are introduced. The notion of partial derivative leads to the equations of tangent planes and to the solutions of optimization problems, generalizing ideas encountered in Calculus I. Finally, the integral is shown to apply to functions defined along a curve in two- and three- space. Appropriate applications are used throughout the course to motivate the need for generalizations.

Introduction (1 class)

Discuss non-trivial examples that clearly exhibit the need to generalize the concepts of derivative and definite integral as they were developed in Calculus I. Encourage members of the class to identify ideas from Calculus I for which generalizations likely will lead to solutions of problems that occur in a three-dimensional world, e.g., slope of a tangent line, critical points, and the definite integral. Applications that might be discussed include average temperature over a region given readings at a discrete set of points, economics-based optimization problems, and work done by a force exerted over a closed path.

The Definite Integral Revisited (9)

- the definite integral: exact values from the Fundamental Theorem of Calculus

- antiderivatives: using substitution, including trigonometric substitution, and integration by parts

- the definite integral: approximate values by using Riemann sums and by the Trapezoidal Rule, with some error analysis.

New applications of the definite integral, or old applications that require new methods for finding antiderivatives or for evaluating integrals of functions that lack simple antiderivatives, should be used to motivate the ideas and techniques in this section. The approach used to introduce new applications should be consistent with that used earlier, that is, using the integral to add up generic pieces of things rather than presenting formulas to be memorized.

Integration by parts and by substitution are included here because they are important in solving differential equations and in other mathematical courses. Other antiderivatives can be found by using tables or a computer algebra system.

Emphasis should be placed on error analysis, including the graph of an appropriate derivative, of the numerical methods used to approximate definite integrals.

Possible applications include arc length (including curves defined by parametric equations), Buffon's needle problem, surface area, numerical integration of tabular data, and Monte Carlo methods to estimate area.

The Integral in $\mathbf{R}^2$ and $\mathbf{R}^3$ (8)

- real valued functions of two and three variables; graphing, level curves
- definitions of double and triple integrals
- integrals over rectangles and boxes
- evaluation of double integrals over regions with curved boundaries

The emphasis in this section is on generalizing the concept of the definite integral to functions of two and three variables. We have consciously restricted the integrals to rectangular coordinates. While there are many nice problems which can be solved more easily using cylindrical and spherical coordinates, time constraints make it advisable to concentrate on the ideas involved in the extension of the concepts to double and triple integrals. In addition, graphing surfaces and drawing level curves pose enough technical problems for most students so that introducing additional complications posed by different coordinate systems seems inadvisable at this point. Computer algebra systems with 3-D graphing capabilities can be very helpful in this section.

Appropriate applications to motivate and give practice in using the new integrals might be taken from probability and economics as well as the usual geometric mathematical applications of finding areas, volumes and centers of mass.

The Derivative in Two and Three Variables (11)

- partial derivatives: definition and geometric motivation
- equation of the tangent plane
- unconstrained optimization: critical points and the Second Derivative Test
- curves described by parametric equations
- Chain Rule
- Extreme Value Theorem revisited

- constrained optimization

- Lagrange multipliers

The geometric analogies to the tangent line should be drawn and used to motivate both the definition of the tangent plane and the method of solution of the optimization problem. The terminology "derivative" should be used, rather than "gradient," for the ordered tuple of partial derivatives to emphasize the analogies. Vector notation should be used only when necessary; the tangent plane is to be expressed as a generalization of the point slope form.

Parametric equations are introduced in the context of constrained optimization, and then only the simpler ones (circles, ellipses, squares, helices, etc.) are studied. Velocity of a point moving along a curve is used to motivate the chain rule.

Those who wish to present a proof of Lagrange's Theorem may want to introduce a bit of vector notation, including dot product, at this point.

The Extreme Value Theorem is stated (but not proved) as justification for constrained optimization.

Constrained optimization is done in two ways: first, comparing values of critical points with boundary values (generalizing the one dimensional case); then using Lagrange multipliers.

Applications should include velocity/speed and several significant applications of Lagrange multipliers (at least one from economics).

Integrations Along Curves (6)

- definition of the Riemann integral of a real function on a curve in $\mathbf{R}^2$ and $\mathbf{R}^3$

- vector fields in $\mathbf{R}^2$ and $\mathbf{R}^3$ and the dot product

- line integrals

- Green's Theorem and path independence

The integral of a (continuous) function on a curve is defined as a straightforward generalization of the Riemann integral of a function of a real variable. Several exercises are worked out and the independence of a parameterization is pointed out. The present goal is to introduce line integral as an example of this integration process. However, the notions of vector fields in $\mathbf{R}^2$ and $\mathbf{R}^3$, (scalar) projection of one vector onto another, and dot product must be introduced first. Line integrals are introduced via application, e.g., perhaps to compute the work done against the wind along certain paths. Green's Theorem is given and could be treated as a two dimensional version of the Fundamental Theorem of Calculus. Path independence, conservative vector fields, and recapturing a primitive are then discussed.

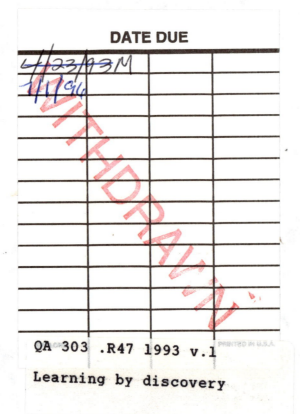